THE GUIDE

INTRODUCED SPECIES

Asterisk (*) indicates species not native to the Columbia Gorge. *See* the "Introduction" (p. 3) for a complete discussion of introduced, native, and endemic plants.

ABUNDANCE

Abundance has been rated on a scale of I through V as follows:
- I: rare
- II: uncommon
- III: moderately abundant
- IV: common
- V: abundant

264*
Rubus discolor   V
(Rubus procerus)
Himalayan Blackberry   (3-8 ft.)
Roadsides, railroads, pastures, and other disturbed open land.
LATE JUNE: Rowland Lake.
LATE JUNE-EARLY JULY: Scenic Highway at Bridal Veil Falls State Park (MP 15.6).
MID JULY: Dalton Point; Larch Mountain Highway.
Rose Family.

HEIGHT OF PLANT

Lists the range in height for each wildflower species.

ABBREVIATION

MP stands for milepost marker. In most cases MP will refer to actual markers, but in some cases (where there are no actual signs) MP refers to the distance along a specific road.

OTHER ABBREVIATIONS:
- BLM: Bureau of Land Management
- elev.: elevation
- DNR: Washington Department of Natural Resources
- ft.: foot or feet
- Hwy.: Highway
- I: Interstate Highway
- in.: inches
- O: Oregon
- SR: Washington State Route
- US: United States Highway
- var.: variety
- W: Washington

FAMILY NAME

Indicates the plant family to which each wildflower belongs. Wildflowers in the "Guide" (pp. 15-272) are grouped according to family name. *See* the five "Guide" contents pages (pp. 16, 72, 128, 178, and 226) for complete listings of family names.

NOTE: This publication is not intended as a passport for entry onto private lands. In the Gorge, private property is still held sacred, particularly since the establishment of the National Scenic Area. Many landowners will permit wildflower hunting and photography on their lands, but do not take chances—*please ask first.*

*Jack
Murdock
Publication
Series
on
the History
of
Science &
Exploration
in
the Pacific
Northwest*

WILDFLOWERS OF THE COLUMBIA GORGE

Written and Photographed by
RUSS JOLLEY

A Comprehensive Field Guide

OREGON HISTORICAL SOCIETY PRESS 1988

Support for this volume was provided in part by funds from the M. J. Murdock Charitable Trust as part of the Jack Murdock Publication Series on the History of Science and Exploration in the Pacific Northwest.

Fold-out map by John Tomlinson and Christine Rains.

Library of Congress Cataloging in Publication Data
Jolley, Russ.
    Wildflowers of the Columbia Gorge.

    (Jack Murdock publication series on the history of science and exploration in the Pacific Northwest)
    Bibliography: p.
    Includes index.
    1. Wild flowers—Columbia River Gorge (Or. and Wash.)—Identification. 2. Columbia River Gorge (Or. and Wash.)—Description and travel—Guide-books. I. Title. II. Series.
QK182.J65   1988   582.13′09795′4   88-1452
ISBN 0-87595-188-0

This volume was designed and produced by the Oregon Historical Society.

Printed in the United States of America.

This book is dedicated to

L M J
E M H
&
S O O

CONTENTS

FOREWORD

SUPERLATIVES are required to describe the Columbia Gorge; ordinary adjectives cannot do justice to this magnificent canyon, carved through a mountain range by western America's largest river. In its looming cliffs geologists can read stories of ancient catastrophes—great outpourings of lava, landslides which filled the river channel, and gigantic floods which carved the mighty cliffs, creating the cascades for which the mountains themselves are named. The Columbia River is a powerful agent of geologic change, holding to its course for millions of years to cut a deep slash in the rising Cascade Range.

In human history, as well, this stretch of river has been the scene of dramatic changes. As the preeminent northwestern gateway between east and west, it was a migration route for countless prehistoric tribes, the pathway of nineteenth-century explorers and pioneers, and now the principal·source of hydropower for a new civilization nurtured by the ancient river. As members of that civilization, we have the right to enjoy the superb bounty of the Gorge, as well as the responsibility to preserve for future generations its store of natural treasures.

The Columbia Gorge is many things to many people. Its beauty can be observed not only on a grand scale of cliffs, waterfalls, and cloud-capped vistas, but also in the world-in-miniature of its animal and plant life. The scenic diversity of the Gorge is, in fact, rivaled by its biological diversity. Abundance of species depends on a rich variety of habitats. The environmental extremes found east and west of the mountains blend in innumerable combinations of sun and shade, heat and cold, moisture and dryness, humus soil and rock, grassland and forest, throughout the seventy-five mile distance separating Troutdale from Maryhill. In the entire Pacific Northwest region, no other area has a concentration of species equal to that found in this one river canyon. To explain and describe the diversity of life is the job of a professional biologist. But anyone can appreciate and enjoy living organisms in their natural settings. The only requirements are patience, a willingness to learn, and the ability to meet nature on its own terms and let it be the teacher.

The richness of the Columbia Gorge as a habitat for plant life can be seen in the diversity of its species of wildflowers. To many persons, the allure of wildflowers is their apparent innocence and independence of humans; they exist without our intervention, yet often display an ordered beauty like plantings in a flower garden. A more realistic view, of course, is that they can indeed be affected by humans, especially when we disturb or alter their native habitats. Some wildflower species are rare and seem to be intolerant of environmental disturbance; others are abundant and hardy, even able to grow on roadsides or in weed-infested meadows. The challenge in studying wildflowers is that there are so many different kinds! How do we go about recognizing all the species and learning their proper names? A very useful tool in this endeavor is a well-illustrated, comprehensive, nontechnical reference book. Surprisingly, there had not been—until now—a field book about wildflowers specifically designed for the Columbia Gorge. This omission is very ably remedied by the book at hand, *Wildflowers of the Columbia Gorge: A Comprehensive Field Guide*.

With some 744 color illustrations, scientific and common names, and notes on abundance, habitat, locations within the Gorge, and blooming seasons for all the species, this work will be of great assistance to anyone who wishes to locate and identify the region's wildflowers. Many of the rare species that characterize the Gorge's unique flora are included, some of them illustrated in color for the first time.

In an era of increasing public concern for the environment, coupled with a sensitivity about the use and conservation of our natural resources, many persons feel a need to become more knowledgeable about the wild plants and animals that share our planet. Wildflowers are of interest not only for their own sake, but also as indicators of the health of the environment. We are indeed fortunate that Russ Jolley has provided this thorough and authoritative guide to the flowers of that magnificent natural feature of the Pacific Northwest, the Columbia Gorge.

KENTON L. CHAMBERS
Oregon State University

ACKNOWLEDGMENTS

NANCY RUSSELL, founder and guiding spirit of Friends of the Columbia Gorge, suggested the idea of this book in 1984. An enthusiastic amateur botanist herself, she saw the need for a field guide to Columbia Gorge wildflowers. Bruce Hamilton, assistant director in charge of publications for the Oregon Historical Society, was receptive to the idea, and after OHS director Thomas Vaughan approved this venture into natural history, the project was underway.

The author is especially indebted to Lois Kemp, a fellow explorer on countless wildflower expeditions into the Gorge. Her sharp eyes and remarkable skill in plant recognition greatly expanded the plant list for the Gorge. That list, *Survey of Wildflowers and Flowering Shrubs of the Columbia Gorge*, by Jolley and Kemp, was published annually by the Native Plant Society of Oregon from 1979 through 1984 and will be updated in the future.

Professor Kenton Chambers, curator of the Herbarium at Oregon State University, identified numerous plant specimens for the author and brought attention to recent changes in plant nomenclature. Professor Bert Brehm, at Reed College, identified several specimens of Desert Parsley (*Lomatium*). Professor Henry Thompson, at UCLA, identified a specimen of Shooting Star (*Dodecatheon*), and Professor Steven Seavey, at Lewis and Clark College, brought attention to nomenclatural changes in the genus *Epilobium*.

The author is also indebted to his longtime friend, Bill Nordstrom, for taking a personal interest in this project and for important advice.

Over the years, many people have helped the author with field observations or identifications and thus have indirectly contributed to this book. Among them are Dick Adlard, Doris Ashby, Louise Godfrey, Liz Handler, Ruth Hansen, Char Corkran, Celeste Holloway, Esther Kennedy, Jim Gamwell, Julie Kierstead, George Lewis, Faith Mackaness, Bob Meinke, Warner Monroe, Steve Nelson, George Pearson, Nancy Russell, Reid Schuller, Jean Siddall, Ruth Strong, Vance Terrall, Krista Thie, Glenn Walthall, and Ann Whitmyer. Two of the above, Liz Handler and Julie Kierstead, also helped the author select photographs for this book.

In all his years of botanizing in the Gorge, the author has never been refused permission to go onto private land in search of plants. Permission was usually granted with an admonition such as, "steer clear of the bull," or "don't forget to shut the gate." The following are among the many landowners who were kind enough to let the author wander over their property: Victor Andres, Claude Berthold, Leon Billette, Dennis Clark, Mrs. Ted Cole, Robert Ezell, L. C. Gove, James Larsen, Luther Olson, Lena Pierce, Tom and John Price, Ted Sauter, Wally Stevenson, Sheldon Struck, H. J. Swager, and John Yeon. Sherry and Larry Kaseberg not only gave the author permission to botanize on Miller Island, but also took time out of their busy ranch schedule to provide the transportation to and from the island.

The author wishes to express his appreciation for the cooperative spirit and professionalism of the entire staff at the Oregon Historical Society Press, publishing arm of the Oregon Historical Society, particularly Bruce Hamilton, Krisell Steingraber, Lori McEldowney, George T. Resch, and Tom Booth. Working with these people has been a pleasure.

Finally, the author is deeply indebted to the Murdock Charitable Trust for a large grant, without which the publication of this book would not have been possible.

Wildflowers of the
Columbia Gorge

INTRODUCTION

IT IS NO WONDER that a marvelous diversity of plant species is packed into this sea-level pass through the Cascade Mountains. Only about fifty miles separate the temperate rain forest at Bonneville Dam (average annual precipitation, seventy-five inches) from the arid grassland at Celilo (average annual precipitation, twelve inches). Furthermore, even as far upriver as The Dalles, the Columbia River is less than one hundred feet above sea level, while elevations reach three and four thousand feet on both sides of the Gorge, adding to the diversity of habitat and thus the diversity of plant species.

These two basic factors, namely, wide ranges in elevation and in precipitation, are responsible for most of the numerous ecological niches that exist in the Gorge. Beyond these, however, the geologic history of the Gorge has generated special environments which provide yet more diversity. For example, mild glaciation during the Ice Age led to the marshes and bogs now found in the upper drainages of the west Gorge. The gigantic floods which marked the end of past ice ages produced vertical basaltic cliffs and waterfalls found mostly in the west Gorge, and vernal ponds, scablands, and dunes in the east Gorge.

While geologic processes over the past 15,000 years were bringing physical changes to the Gorge, radical changes in climate were inducing vast migrations of plants into and through the Gorge. Leroy Detling, a paleobotanist who was curator of the University of Oregon Herbarium for over thirty years, postulated that the marked climatic changes from very cold, to warm and dry, to the present cool and moist conditions, have caused plant migrations, first from the north and from high elevations to the floor of the Gorge, and then later from the south along the Cascade Range and from the Rogue River Valley. Many of these migrants found suitable niches in the Gorge, niches which they still occupy.

If it is true that the number of native plant species found in a region is directly related to the number of different types of habitat available, then it is not surprising that more than eight hundred species of native wildflowers and flowering shrubs have been found in the Gorge. This number is almost a quarter of the total native wildflower species count for the entire state of Oregon.

Finally, the most recent influence on Gorge flora was the arrival of the settlers. They brought with them their garden flowers and grain seed supplies, the latter all too often contaminated with seeds of European weeds. Thus, accidentally or on purpose, a multitude of introduced plants are now well established in the Gorge, in many cases displacing the native flora.

Vegetation Zones

The Columbia Gorge can be divided into five major vegetation zones (see fold-out map), including three coniferous forest zones (Western Hemlock Zone, Douglas-fir Zone, and Silver Fir Zone), a Pine-oak Woodland Zone, and a Shrub-grassland Zone.

The Western Hemlock Zone extends from the western boundary of the Gorge to the vicinity of Wind Mountain, about thirty-two miles. In this zone, Western Hemlock (*Tsuga heterophylla*) is the tree species that dominates the climax forest stands, i.e. stands that are self-perpetuating. Actually, because of past logging and forest fires, especially the great Columbia Fire of 1902 which ravaged forests on both sides of the river, there is not much climax forest left in the west end of the Gorge. At present, the dominant species in these immature forests of the Western Hemlock Zone is Douglas-fir (*Pseudotsuga menziesii*). In a matter of centuries, however, the Douglas-fir will be entirely replaced by Western Hemlock and other shade-tolerant tree species, barring further disturbance by fire, logging, or other catastrophe.

The Silver Fir Zone is found in scattered areas of the west Gorge that rise above three thousand feet in elevation. In this zone, Pacific Silver Fir (*Abies amabilis*) is the climax dominant tree species. In the early stages before the climax condition is achieved, Douglas-fir, Noble Fir, and Western Hemlock are important components of forests in the Silver Fir Zone.

The Douglas-fir Zone generally lies just east of the crest of the Cascade Mountains. It includes the drier forests from Wind Mountain east to the vicinity of Bingen, a distance of about fifteen miles. The climax forest of this zone is open and relatively brush free, dominated by Douglas-fir trees rather than Western Hemlock.

The Pine-oak Woodland Zone extends from Bingen almost to The Dalles, some ten or twelve miles. Two tree species, Ponderosa Pine (*Pinus ponderosa*) and Oregon White Oak (*Quercus garryana*) are co-dominants, along with grassy openings on south-facing slopes and Douglas-fir on north-facing slopes.

The Shrub-grassland Zone extends from The Dalles eastward. This is primarily open grassland, with scattered shrubs such as Bitter Brush (*Purshia tridentata*), Mock Orange (*Philadelphus lewisii*), and Rabbit Brush (*Chrysothamnus nauseosus*).

Obviously, the actual boundaries between vegetation zones are not so sharp and smooth as shown on the fold-out map. For example, there are narrow stringers of pine-oak woodland extending westward along the Columbia River to Dog Mountain on the north side, and to Starvation Creek on the south side. A narrow band of shrub-grassland also extends as far west as the Klickitat River.

Columbia Gorge Endemics

In a class by themselves are the fifteen Columbia Gorge endemics, wildflowers that are found only in the Columbia Gorge and vicinity. Seven of these are typically cliff-dwellers; five grow on gentle slopes and flats in pine-oak woodlands; two are found in shrub-grasslands; and one is known to grow only at the eastern tip of Miller Island. (The number listed in parentheses following each flower name is the picture number and refers the reader to a detailed description contained in the Guide section.)

The cliff-dwelling endemics include *Erigeron oreganus* (674), invariably found under overhanging basalt cliffs; *Sullivantia oregana* (224), which grows on cliffs within reach of the spray from waterfalls; *Douglasia laevigata var. laevigata* (436), normally found on vertical cliffs throughout the west Gorge; *Hieracium longiberbe* (616), at home on both cliffs and rocky slopes; *Erigeron howellii* (672), a resident of shady cliffs and steep north-facing slopes in the west Gorge; *Synthyris stellata* (567), also found on cliffs and north-facing slopes; and *Penstemon barrettiae* (554), which grows on cliffs and talus slopes on both sides of the Columbia River in the vicinity of Mosier, and also near Bonneville Dam.

Endemic species that grow in the pine-oak woodlands of the Gorge include *Lomatium suksdorfii* (398), usually found at middle to high elevations; *Lomatium columbianum* (388), commonly found at low elevations; and *Dodecatheon poeticum* (434), also commonly found at low elevations, although its range slightly exceeds the boundaries of the Pine-oak Woodland Zone on both east and west. *Lupinus latifolius var. thompsonianus* (299) is typically found at low elevations. Its range extends sporadically to the east end of the Gorge, but the largest populations are found in pine-oak woodlands. *Astragalus hoodianus* (279) is found at both low and high elevations. A black-hairy form grows in pine-oak woodlands between Hood River and The Dalles, while a white-hairy form grows in grasslands along SR-14 near the junction with The Dalles Bridge Road.

Two other Columbia Gorge endemics are found in the Shrub-grassland Zone. The center of the known range of *Ranunculus reconditus* (147) is near The Dalles, but all the populations of this plant are located at high elevations on both sides of the Columbia River, near the boundaries of the Gorge shown on the map. *Lomatium laevigatum* (391) finds a home on cliffs and rocky benches at low elevations from The Dalles eastward.

Artemisia campestris var. wormskioldii (not pictured) once grew on rocky banks of the Columbia River throughout the east Gorge. However, most of the suitable habitat was submerged behind Bonneville Dam and The Dalles Dam. The one known extant site lies a few feet above the water at the east end of Miller Island, a total population of fewer than twenty plants.

Defining the Gorge

The eastern and western limits of the Columbia Gorge are here understood to be the Sandy River on the west and the town of Biggs Junction on the east. From the Sandy River to Biggs Junction is a span of about seventy-five miles as the crow flies, or eighty-eight miles as measured along the Columbia River.

On the map, the vertical lines numbered 0 to 19 are spaced at intervals of five minutes (5') of longitude, or about four miles. These lines are intended to provide a reference grid for locating places in the Gorge, as are the prefixes O and W, which indicate Oregon and Washington sides, respectively. For example, Larch Mountain has the notation O/3.9, while that of Stacker Butte is W/15.8. Always read to the right of the latitude lines. Each entry in the Place Names section has its appropriate notation.

As used here, the term *west Gorge* refers to that portion of the Gorge lying west of the White Salmon River, while *east Gorge* refers to that portion lying east of the river. In a remarkable number of cases, the White Salmon River does, in fact, represent the eastern or western range limit for Gorge plants. Many plant species, however, especially those characteristic of the Douglas-fir Zone, have ranges that lie in what might be called the *middle Gorge,* roughly the stretch of land between the Wind River and the Klickitat River.

The northern and southern limits of the Gorge are not so easily defined. However, it seems logical to consider the Gorge as having the shape of a trough whose rims on either side are the ridge crests at the heads of the creeks that flow into the Gorge. For example, Rock Creek and Major Creek on the north side, and Eagle Creek and Mosier Creek on the south side, are included as part of the Gorge. The northern boundary of the Gorge could thus be outlined by linking high elevation points from Silver Star Mountain east to Haystack Butte. On the south, a similar series of peaks includes Larch Mountain, Indian Mountain, and Mt. Defiance. Using these northern and southern limits, the Gorge varies from five to twenty miles in width.

Criteria for Wildflowers Selected

As originally conceived, this book was to present all the native wildflowers of the Gorge, and some of the introduced flowers. The term "wildflower" turns out to be vague, however. Whole families of flowering plants fall outside the commonly accepted concept of wildflowers. Using an arbitrary criterion, therefore, plants whose flowers are so inconspicuous that they generally would not be identified as flowers from a distance of a few feet were excluded.

Excluded on this basis were families such as grasses, sedges, goosefoot, pigweed, and many other flowering herbaceous plants, shrubs, and trees. In fact, most trees were excluded, even the rather conspicuous willows and tassel-bearing trees such as alders and maples. Included, on the other hand, were some borderline species, such as *Limosella aquatica* (533), whose tiny white flowers are just visible against a background of mud, and *Pachistima myrsinites* (332), whose maroon flowers were judged to qualify, but just barely. *Orthocarpus pusillus* (548) has microscopic flowers too small to qualify, but despite the rule, it was included. About thirty native wildflower species, almost all of them rare in the Gorge, would have qualified, but are not represented here because the author, for a variety of reasons, did not get photographs of them.

With the introduced species, the intent was not to be all-inclusive, but rather to select those species that have become most characteristic of the Gorge, and those that are most showy. Most of the common introduced garden and lawn weeds, such as dandelions and clovers, were not included. Also omitted was Scotch Broom, a scourge which is spreading on roadsides and open areas in the west Gorge.

Plant Names

Every plant has a scientific name consisting of two principal parts, a generic name (genus) and a specific name (species). For example, Western Buttercup, a member of the large Buttercup Family, has the scientific name, *Ranunculus occidentalis*, the first name, *Ranunculus*, denoting the genus and the second name, *occidentalis*, denoting the species. Some species are further subdivided into two or more different varieties, each usually characteristic of a separate geographic area, much as the different races of *Homo sapiens* are found in different parts of the world. Thus, *Allium douglasii var. nevii*, a member of the Lily Family, is found in the Columbia Gorge, while three other varieties of *Allium douglasii* are found in other well-marked areas of the Pacific Northwest.

Most of the scientific names used here are those found in *Flora of the Pacific Northwest*, by Hitchcock and Cronquist. In a few cases, however, other valid names have been used. Also, some recently accepted changes in nomenclature have been incorporated.

Common names are a different matter, however. Many plant species have two or more common names, sometimes as many as half-a-dozen. In contrast, some native wildflowers, including several showy species, appear to have no common name. When alternative common names were available, choice went to the Indian name, or to the most descriptive name.

The plant families are arranged here in the same order as they appear in *Flora of the Pacific Northwest*, i.e., the Englerian sequence. Within each family, the genera and species are listed in alphabetical order. Introduced species are indicated by an asterisk (*) following the picture number.

The size of each pictured plant is given in inches or feet, except in the cases of certain water plants and vines. Also, the relative abundance of each species in the Gorge is estimated on a scale of I (rare) to V (abundant). Abundance Class I includes plants that are found in only one or a few places in the Gorge, and only in limited numbers at these sites. At the other end of the scale, Class V includes species that are abundant over a wide range in the Gorge.

While this book is designed as a field guide to the plants pictured here, it can also be used to help identify unknown plants encountered in the field. One simply searches through the book to find the picture that most resembles the unknown plant. This is a slow process, at least for the beginner, but it gets easier as one becomes more familiar with the plant families. Bear in mind, that many introduced wildflowers and inconspicuous natives are not represented here. For accurate identification of unknown plants, consult *Flora of the Pacific Northwest*, referred to previously.

Wildflower Locations

For each species, the caption contains a brief description of its typical habitat in the Gorge, and reference is made to one or more places where the plant can be found. Most of the places referred to were chosen for ease of access and relative immunity to visitor impact. Admittedly, this information could expose some plants to predation by plant collectors ("diggers"), but this will probably not become a serious problem. The heaviest foreseeable impact from wildflower lovers could not compare in severity with that from cattle-grazing, logging, or abuse by off-road vehicles. Since most of the sites referred to are located near roads, some sites will be vulnerable to highway department activities (widening, filling, ditch maintenance, herbicide spraying, mowing, etc.). In fact, it would be remarkable luck if some of the sites are not altered or destroyed in coming years.

In general, the most sensitive and pristine areas are not mentioned. As a consequence, specific site references are omitted for a few species. In the west Gorge, there are waterfalls, marshes, and subalpine ridges which cannot be reached by road or trail. In the east Gorge, there are grassy benches bounded by cliffs, where cattle have never grazed.

Some of these places are still so unspoiled that no introduced plants have been found there.

Blooming Times

Based on more than ten years of observation, an estimate was made of the time of the year when each plant would be in bloom at particular locations. Many species have long blooming seasons, while others have short seasons. For these short-season plants, it is harder to be confident of the suggested blooming time, especially since general weather conditions can sometimes shift the entire blooming season by two or even three weeks. In an attempt to compensate for this uncertainty, several sites are usually mentioned.

Hazards of Wildflower Hunting

PARKING   Parking along Gorge highways can be hazardous unless certain simple rules are observed. First, always park *off the pavement*, except where a paved turnout is provided for parking. Second, take great care when re-entering the traffic lane from the parking spot.

PRIVATE LANDS   This book is not intended as a passport for entry onto private lands. In the Gorge, private property is still held sacred, particularly since establishment of the National Scenic Area. Many landowners will permit wildflower hunting and photography on their lands, but do not take chances—ask first.

WIND   This can be a hazard to wildflower photography. The best condition is to have no wind at all, of course, but calm in the Gorge is not the norm. With an east wind, the best place to be is the east Gorge, especially The Dalles area and eastward. Often, when there is a forty mile-per-hour gale from the east at Cape Horn, conditions are calm at The Dalles. With a west wind, the best place to be is anywhere west of Bonneville Dam. The flowers at Cape Horn can be standing almost still while windsurfers are having a great time at Hood River.

POISON OAK   The reader who is not sensitive to Poison Oak can skip this section. It is only an occasional plant west of Cascade Locks, but between there and The Dalles, it is extremely common at low elevations. It can be a low ground cover in shaded situations or a robust shrub on sunny open slopes. There are reports of a protective cream which those who are Poison Oak-sensitive can apply before exposure, but the best defense is avoidance. Walking along roads and trails should be safe, depending on the level of one's sensitivity, but cross-country travel in pine-oak woodlands can often carry a definite risk of exposure.

TICKS   These are a nuisance of springtime, particularly in March and April, and mostly between Cascade Locks and The Dalles. Mossy places may invite one to sit down a spell, but it is better to resist that temptation. Brushy places regularly grazed by deer are likely spots for picking up a tick. For what it may be worth, the author's anti-tick strategy consists of: (1) Eating a clove of garlic (diced, on buttered toast) for breakfast; one can get to like it. (2) Spraying or dusting socks and pantlegs with a commercial tick repellant. (3) Now and then checking pantlegs for the little creatures, especially when traveling through brushy areas. (4) Back at home, checking skin and all clothing for ticks.

If a tick is attached, just pull it off. Some people touch the tick with a little bit of fingernail polish remover to get it to back out. A fear of ticks should not prevent anyone from hiking in the east Gorge. To reduce the chances of picking up a friendly tick, stay on the roads and trails during early spring, and do not sit in strange places.

RATTLESNAKES   Snakes are the most overrated hazard of the Gorge, but they have to be kept in mind. Yes, there are rattlers, up to three feet long, from Wind Mountain eastward. They are more common on the drier and more open north side of the Columbia River. However, in twenty years of intensive bushwacking in the Gorge, the author has seen only eleven rattlesnakes, all but one of them going the other way. Only one, trapped, coiled to defend itself. If frightened by sounds transmitted through the ground, they will try to escape to safe cover, usually under rocks. Stamping the feet while walking and tapping a hiking staff apparently does the trick.

Author's Purpose

This book is intended as a field guide to wildflowers of the Columbia Gorge—to help the reader find and identify flowers when they are blooming. Its further purpose is to stimulate a wider interest and appreciation of the Gorge and its native plants, and thereby secure increased protection for them.

The key to protection of native plants is preservation of their habitat. The areas of undisturbed habitat still remaining in the Gorge are treasures of inestimable value. They should be treated with the greatest care. The most serious and irreversible threats to plant habitat are those involving land disturbance, especially conversion of previously undisturbed land to roads, gravel pits, croplands, orchards, and residential and industrial developments. Wherever bulldozers and tractors alter the face of the land, we can say goodbye to most of the native plants.

While land disturbance is the major threat to our native plants, there are others. Herbicides used by highway departments take their toll of wildflowers along the roadsides. Overgrazing destroys a host of beautiful flowers, including Northwest Balsamroot (723) and Green-banded Mariposa Lilies (17). Aggressive weeds such as the Himalayan Blackberry, English Ivy, and Scotch Broom—all of them introduced—are usurping native plant habitat throughout the Gorge.

Even wildflower lovers can be a threat. Picking, trampling, digging, and collecting all impact wildflower populations. Each of us must take responsibility for leaving the natural treasures of the Gorge as beautiful and natural as we found them.

Serendipity

The classic example of serendipity was, of course, Columbus. He was looking for a new way to China, but instead found the New World. In its own way, wildflower hunting, too, is a serendipitous pursuit. Inevitably, the careful explorer will find unexpected plants. Certainly, there are wildflower species growing in the Gorge that have not yet been reported, and the reader is invited to join the happy search.

SECTION ONE

1
Alisma gramineum   I
var. angustissimum
Narrow-leaf   (6-18 in.)
Water Plantain
Muddy shores of ponds and streams.
MID JUNE: The Dalles Bridge Ponds.
Plant does not appear in dry years.
Water Plantain Family.

2
Alisma plantago-aquatica   III
American Water Plantain   (1-2.5 ft.)
Ponds in the east Gorge; Columbia
River bottomlands throughout the
Gorge.
EARLY JULY: Tom McCall Nature
Preserve; Crates Point Wildlife Area;
vernal ponds at Horsethief Lake State
Park.
MID JULY: Stevens Pond.
Water Plantain Family.

3
Machaerocarpus   I
californicus
Fringed Water Plantain   (1-2 ft.)
Vernal ponds near The Dalles.
LATE JUNE: pond on BLM land near
The Dalles Mountain Road; ponds in
and near Horsethief Lake State Park.
Water Plantain Family.

4
Sagittaria cuneata   II
Arum-leaf Arrowhead,   (6-18 in.)
Wapato
Columbia River shores and bottom-
lands; ponds from the Sandy River to
The Dalles area.
LATE JULY: Tom McCall Nature Pre-
serve.
EARLY SEPT: Columbia River shore
east of The Dalles Riverside Park.
Water Plantain Family.

5
Sagittaria latifolia   III
Broad-leaf Arrowhead,   (1-3 ft.)
Wapato
Columbia River shores and bottom-
lands in the west Gorge.
MID-LATE AUG: Mirror Lake and
Young Creek in Rooster Rock State
Park; Franz Lake.
Water Plantain Family.

BUR REED FAMILY

6
Sparganium emersum   II
var. emersum
Simple-stem Bur Reed   (8-18 in.)
Ponds and marshes at low to middle
elevations; Columbia River bottom-
lands.
LATE JULY-EARLY AUG: Tom McCall
Nature Preserve; Beacon Rock Pond.
Bur Reed Family.

7
Lysichitum americanum   III
Skunk-cabbage   (1.5-3.5 ft.)
Swampy ground, ditches, and slow
streams at all elevations in the west
Gorge.
EARLY APRIL: Latourell Falls; Eagle
Creek Forest Camp; Beacon Rock
Pond.
LATE APRIL: Palmer Mill Road; Belle
Center Road.
Arum Family.

8
Allium acuminatum   V
Taper-tip Onion   (6-12 in.)
Dry open areas throughout the Gorge,
but chiefly in the east Gorge.
LATE MAY-EARLY JUNE: Old Highway
at Catherine Creek.
EARLY-MID JUNE: Signal Rock; Dog
Creek Falls; Tom McCall Nature Pre-
serve south of US-30.
Lily Family.

9
Allium amplectens   I
Narrow-leaf Onion   (6-12 in.)
Dry rocky streambeds at low elevations
between Bingen and Dallesport.
LATE MAY-EARLY JUNE: Old Highway
at dry streambeds near Catherine
Creek.
Lily Family.

10
Allium cernuum   II
Nodding Onion   (6-18 in.)
Moist soil in open places at all eleva-
tions in the west Gorge.
MID JULY: Scenic Highway near
Crown Point and also just east of
Horsetail Falls; high on the Hamilton
Mountain Trail; top of Angels Rest.
Lily Family.

11
Allium crenulatum   II
Scalloped Onion   (2-6 in.)
Gravelly open places near the crest of
the Cascade Mountains; thus far
found only on the Washington side,
but locally abundant there.
LATE MAY-EARLY JUNE: Grassy Knoll
Trailhead.
MID JUNE: top of Big Huckleberry
Mountain.
Lily Family.

12
Allium douglasii var. nevii   II
Douglas' Onion   (6-16 in.)
Moist open grasslands in the east
Gorge.
EARLY MAY: private land off US-30
near MP 4.7 east of Mosier.
MID MAY: top of The Dalles Mountain
Road; Dry Creek Road near the power-
line crossing.
Lily Family.

13
Brodiaea congesta   V
Ball-head Cluster Lily,   (1.5-2.5 ft.)
Ookow
Grassy open or lightly wooded slopes
as far east as The Dalles.
MID MAY: Rock Creek Road.
LATE MAY-EARLY JUNE: Dog Creek
Falls.
EARLY JUNE: Tom McCall Nature
Preserve; Dog Mountain Trail.
MID JUNE: Hood River Mountain
Meadow.
Lily Family.

14
Brodiaea coronaria   III
Harvest Brodiaea   (4-14 in.)
Dry grassy areas as far east as Horse-
thief Lake State Park, but chiefly
between Dog Mountain and Lyle.
MID-LATE JUNE: Old Highway near
Catherine Creek; SR-14 at the
Broughton Lumber Mill near MP 62.
Lily Family.

15
Brodiaea howellii   V
Bicolored Cluster Lily   (1-2 ft.)
Grassy open slopes throughout the
Gorge, chiefly east of Dog Mountain.
MID APRIL: SR-14 east of Lyle Tunnel
(MP 77-81).
LATE APRIL-EARLY MAY: SR-14 near
Tunnel #1.
MID MAY: top of The Dalles Mountain
Road.
Lily Family.

16
Brodiaea hyacinthina   III
Hyacinth Cluster Lily   (1-2.5 ft.)
Moist open grasslands as far east as
Horsethief Lake State Park.
EARLY JUNE: SR-14 at the base of
Wind Mountain and east of Lyle
Tunnel (MP 77-81); Old Highway at
the gravel pit just east of Major Creek.
MID JUNE: Bridal Veil Falls State Park.
Lily Family.

17
Calochortus macrocarpus   II
Green-banded Mariposa Lily   (1-2 ft.)
Dry, open or lightly wooded slopes in
the east Gorge, often near poison oak.
LATE JUNE: Mayer State Park; Tom
McCall Nature Preserve; Memaloose
State Park; base of cliffs at Horsethief
Lake State Park.
Lily Family.

18
Calochortus subalpinus   II
Cascade Mariposa Lily,   (4-10 in.)
Cat's-ear Lily
High open ridges in the west Gorge.
MID-LATE JUNE: Grassy Knoll
Trailhead; top of Big Huckleberry
Mountain; top of Monte Carlo.
Lily Family.

19
Camassia leichtlinii   I
Great Camas   (1-2.5 ft.)
Ditches and wet meadows in the west
and middle Gorge.
MID MAY: ditches beside SR-14
between Washougal and the Clark-
Skamania County Line; woods below
SR-14 near MP 19.5; wet meadows on
the Major Creek Plateau.
Lily Family.

20
Camassia quamash   IV
Common Camas   (1-2 ft.)
Wet meadows, ditches, and moist
banks as far east as Horsethief Lake
State Park.
MID APRIL: SR-14 near Carson Depot
Road; Old Highway near Catherine
Creek.
LATE APRIL-EARLY MAY: visible from
I-84 at Bridal Veil; Bridal Veil Falls State
Park.
Lily Family.

21
Clintonia uniflora   IV
Bead Lily   (4-8 in.)
Coniferous woods in the west Gorge,
mostly at middle to high elevations.
LATE JUNE: Monte Carlo Trail.
EARLY JULY: Larch Mountain High-
way; Palmer Mill Road.
MID JULY: Spring Camp Road; top of
Larch Mountain.
Lily Family.

22
Disporum hookeri V
var. oreganum
Fairy Bells (1-2 ft.)
Coniferous forests in the west and
middle Gorge, commonly at low
elevations.
LATE APRIL: Latourell Falls; Multno-
mah Falls; Beacon Rock.
Lily Family.

23
Disporum smithii II
Fairy Lanterns (1-2 ft.)
Coniferous woods at the west end of
the Gorge.
EARLY-MID JUNE: Larch Mountain
Highway from MP 3.5 to the Mt.
Hood National Forest boundary.
Lily Family.

24
Erythronium grandiflorum IV
Glacier Lily (6-14 in.)
Open or lightly wooded slopes as far
east as The Dalles, usually near oaks.
LATE MAR: Memaloose Rest Area;
Old US-30 west of Mosier; US-30
between Mosier and Mayer State
Park; Old Highway.
LATE MAY-EARLY JUNE: Monte Carlo
Trail.
Lily Family.

25

Erythronium montanum II
Avalanche Lily (6-14 in.)
Subalpine forests and meadows above
2,800 ft. elev. in the west Gorge.
MID MAY: Monte Carlo Trail.
EARLY-MID JUNE: top of Larch Mountain.
LATE JUNE: Indian Mountain.
Lily Family.

26

Erythronium oregonum I
Fawn Lily (6-16 in.)
Moist woods at low elevations in the
far west end of the Gorge.
LATE MAR-EARLY APRIL: woods
below SR-14 east of Washougal (MP
19.5-20.5).
Lily Family.

27

Fritillaria lanceolata IV
Chocolate Lily (1-2.5 ft.)
Open woods, usually near oaks, at low
to middle elevations as far east as The
Dalles.
EARLY APRIL: Stanley Rock; Campbell
Creek; backroads south of Mosier.
MID APRIL: Eagle Creek Overlook
Picnic Area; Signal Rock.
LATE APRIL: McCord Creek Falls Trail.
MAY: Hamilton Mountain Trail.
Lily Family.

28
Fritillaria pudica   IV
Yellow Bells   (5-10 in.)
Open grasslands at all elevations east
of Dog Mountain.
MID MAR: SR-14 east of Lyle Tunnel
(MP 77-81); Tom McCall Nature
Preserve.
EARLY APRIL: top of Vensel Road;
Hood River Mountain Meadow.
Lily Family.

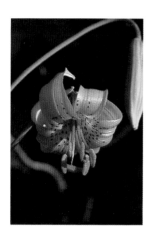

29
Lilium columbianum   IV
Tiger Lily   (2-5 ft.)
Coniferous woods and forest openings
in the west and middle Gorge.
MID JUNE: Scenic Highway at Bridal
Veil, Bridal Veil Falls State Park, Mult-
nomah Falls, and Ainsworth State
Park.
MID-LATE JULY: Larch Mountain
Highway near the snowgate (MP 10);
Indian Mountain.
Lily Family.

30
Maianthemum dilatatum   III
False Lily-of-the-valley   (5-12 in.)
Moist woods at all elevations in the
west Gorge.
MID MAY: Scenic Highway at Bridal
Veil Creek and Ainsworth State Park;
Ridge Trail in Rooster Rock State Park.
EARLY JULY: Spring Camp Road; top
of Larch Mountain.
Lily Family.

31
Smilacina racemosa   V
Western Solomon Plume   (1-3 ft.)
Open woods at all elevations in the
west and middle Gorge.
EARLY MAY: SR-14 from Washougal
to Beacon Rock.
MID MAY: Scenic Highway from
Crown Point to Ainsworth State Park.
EARLY JULY: top of Larch Mountain.
Lily Family.

32
Smilacina stellata   IV
Starry Solomon Plume   (6-18 in.)
Moist woods in the west and middle
Gorge, chiefly at middle to high eleva-
tions.
EARLY MAY: Ridge Trail in Rooster
Rock State Park.
LATE MAY-EARLY JUNE: Grassy Knoll
Trailhead; Larch Mountain Highway
east of MP 3.5.
LATE JUNE-EARLY JULY: top of Larch
Mountain.
Lily Family.

33
Stenanthium occidentale   III
Bronze Bells   (8-16 in.)
Cliffs, talus slopes, and rocky stream-
banks in the west Gorge.
EARLY JUNE: Latourell Falls; Multno-
mah Falls; Beacon Rock.
Lily Family.

34
Streptopus amplexifolius III
Clasping-leaf (1.5-3.5 ft.)
Twisted Stalk
Streambanks and moist forests in the
west Gorge.
MID-LATE MAY: Latourell Falls;
Haines Road Bridge over Latourell
Creek; Scenic Highway just east of
Crown Point; Gorge Trail at Moffett
Creek.
Lily Family.

35
Streptopus roseus II
Rosy Twisted Stalk (6-12 in.)
Moist woods at middle to high eleva-
tions in the west half of the Gorge.
LATE JUNE: North Lake Trailhead.
Lily Family.

36
Tofieldia glutinosa III
var. brevistyla
Western False Asphodel (1-2.5 ft.)
Wet meadows and boggy areas at
middle to high elevations in the west
Gorge.
EARLY JULY: north and west shores of
Rainy Lake.
Lily Family.

37
Trillium ovatum V
Western Wake-robin (6-16 in.)
Common in moist woods in the west
and middle Gorge.
LATE MAR: Scenic Highway near
Bridal Veil and at Ainsworth State
Park; Lower Tanner Creek Road.
EARLY JUNE: top of Larch Mountain.
Lily Family.

38
Veratrum californicum II
var. caudatum
White False Hellebore (4-7 ft.)
Marshes, wet meadows, and slow
streams in the west and middle Gorge.
LATE JUNE: meadows on the Major
Creek Plateau; Laws Corner.
Lily Family.

39
Veratrum insolitum I
Siskiyou False Hellebore (4-8 ft.)
Dry open woods and brush in the
middle Gorge.
MID JULY: Nestor Peak Road (Road
N-1000), especially near Buck Creek
Trailhead #1.
Lily Family.

40
Veratrum viride   II
Green False Hellebore   (3-6 ft.)
Wet meadows and moist open woods
at middle to high elevations in the west
Gorge.
LATE AUG: near Indian Springs.
Lily Family.

41
Xerophyllum tenax   IV
Bear-grass   (2-4 ft.)
Open woods and clearings, generally
at higher elevations in the west Gorge.
EARLY JULY: top of Larch Mountain;
Indian Springs; Rainy Lake.
Lily Family.

42
Zigadenus paniculatus   II
Panicled Death-camas   (1-2 ft.)
Dry open slopes at low elevations in
the east Gorge.
EARLY APRIL: SR-14 east of Lyle
Tunnel (MP 77-81); Horsethief Butte;
The Dalles Mountain Road.
Lily Family.

43
Zigadenus venenosus   IV
var. venenosus
Meadow Death-camas   (6-18 in.)
Grassy open slopes at low to middle
elevations, generally between Hamil-
ton Mountain and Horsethief Butte.
LATE APRIL-EARLY MAY: Old
Highway; Horsethief Butte.
MID MAY: base of Wind Mountain;
SR-14 at Carson Depot Road
EARLY-MID JUNE: Hood River Moun-
tain Meadow.
Lily Family.

IRIS FAMILY ───────────

44
Iris missouriensis   I
Western Blue Flag   (1-2 ft.)
Vernally moist meadows and stream-
banks in the east Gorge.
MID MAY: Horsethief Butte Trailhead.
Iris Family.

45*
Iris pseudacorus   II
Yellow Flag   (2-3 ft.)
Low elevation streambanks and lake
margins from Bridal Veil to Horsethief
Butte.
MID-LATE MAY: Grant Lake.
Iris Family.

46
Iris tenax III
Oregon Flag (8-16 in.)
Forest openings at low to middle
elevations in the west Gorge.
JUNE: Larch Mountain Corridor;
Brower Road; bluff west of Bridal Veil
Falls; Gibson Road.
Iris Family.

47
Sisyrinchium douglasii V
var. douglasii
Grass Widow (6-12 in.)
Open grassy areas in the middle and
east Gorge, particularly in low elevation
areas between Dog Mountain and
Dallesport.
EARLY MAR: SR-14 east of Lyle Tunnel
(MP 77-81).
MID MAR: Old Highway; Memaloose
Viewpoint; Stanley Rock.
Iris Family.

48
Sisyrinchium douglasii II
var. inflatum
Grass Widow (6-12 in.)
Open grassy areas in the east Gorge,
especially high elevation areas east of
The Dalles.
MID-LATE MAR: top of The Dalles
Mountain Road.
Iris Family.

49
Sisyrinchium idahoense   II
(Sisyrinchium angustifolium)
Blue-eyed Grass   (6-14 in.)
Moist open areas at scattered locations,
chiefly in the middle and east Gorge.
LATE MAY: SR-14 at Carson Depot
Road and at Grant Lake.
EARLY JUNE: meadows on the Major
Creek Plateau; moist grasslands at the
north edge of The Dalles Airport.
MID JUNE: top of Angels Rest.
Iris Family.

ORCHID FAMILY

50
Calypso bulbosa   III
Deer's-head Orchid,   (4-8 in.)
Fairy Slipper
Shady coniferous forests at low to
middle elevations, almost as far east as
The Dalles.
MID-LATE APRIL: Eagle Creek Forest
Camp; Wyeth Trailhead; Wygant Trail
west of the Mitchell Point Picnic Area.
Orchid Family.

51
Corallorhiza maculata   III
Spotted Coral Root   (6-16 in.)
Coniferous woods at all elevations in
the west and middle Gorge.
MID-LATE JUNE: Major Creek Plateau;
Nestor Peak Road (Road N-1000).
EARLY-MID JULY: Pacific Crest Trail
near Big Huckleberry Mountain.
Orchid Family.

52
Corallorhiza mertensiana   III
Mertens' Coral Root   (8-18 in.)
Coniferous woods at middle to high
elevations in the west and middle
Gorge.
JULY: woods along Larch Mountain
Highway at the Mt. Hood National
Forest boundary; Spring Camp Road.
Orchid Family.

53
Corallorhiza striata   II
Striped Coral Root   (6-16 in.)
Coniferous and deciduous woods and
brush at low to middle elevations
between Cascade Locks and The
Dalles.
MID MAY: top of Vensel Road; near
Wasco Butte Lookout.
MID JUNE: Dog Mountain Trail.
Orchid Family.

54
Corallorhiza trifida   I
Northern Coral Root   (8-12 in.)
High elevation woods just west of Mt.
Defiance.
EARLY-MID JULY: near North Lake
Trailhead.
Orchid Family.

55

Cypripedium montanum   I
Mountain Lady's Slipper   (1-2 ft.)
Woods and brush at middle to high
elevations between Dog Mountain
and The Dalles.
LATE MAY-EARLY JUNE: Major Creek
Plateau; DNR Forest; higher elevations
between Mosier and The Dalles.
Orchid Family.

56

Eburophyton austiniae   II
(Cephalanthera austiniae)
Phantom Orchid   (8-16 in.)
Shady coniferous woods with little
underbrush at low to middle elevations
in the west and middle Gorge.
MID-LATE JUNE: Dog Mountain Trail.
LATE JUNE-EARLY JULY: backroads on
the Major Creek Plateau.
Orchid Family.

57

Epipactis gigantea   I
Giant Helleborine   (1-2 ft.)
Low elevation streambanks at scattered
locations in the middle and east Gorge.
EARLY JULY: Wind River shore above
the mouth of the Little Wind River.
Orchid Family.

58
Goodyera oblongifolia   II
Rattlesnake Plantain   (8-16 in.)
Coniferous woods with little under-
brush at all elevations in the west and
middle Gorge.
MID-LATE JULY: Fort Cascades Historic
Site.
EARLY-MID SEPT: top of Larch Moun-
tain.
Orchid Family.

59
Habenaria dilatata   II
var. dilatata
(Platanthera dilatata)
White Bog Orchid   (1-3 ft.)
Wet meadows and boggy lake margins
in the west and middle Gorge.
LATE JUNE: meadows on the Major
Creek Plateau.
LATE JULY: Rainy Lake.
Orchid Family.

60
Habenaria elegans   II
(Platanthera elegans)
Long-spurred Rein Orchid   (1-2 ft.)
Open woods at low to middle eleva-
tions between Bonneville Dam and
The Dalles.
LATE JUNE-EARLY JULY: Tom McCall
Nature Preserve.
MID JULY: Fort Cascades Historic Site.
Orchid Family.

61
Habenaria saccata   III
(Platanthera saccata)
Slender Bog Orchid   (1-2.5 ft.)
Wet meadows, lake margins, and
springs at middle to high elevations in
the west Gorge.
MID JULY: north and west shores of
Rainy Lake; North Lake Trailhead.
Orchid Family.

62
Habenaria sparsiflora   I
(Platanthera sparsiflora)
Few-flowered Bog Orchid   (1-2 ft.)
MID-JULY: wet meadows at middle
elevations west of Bonneville Dam.
Orchid Family.

63
Habenaria unalascensis   II
(Platanthera unalascensis)
Short-spurred Rein Orchid   (8-18 in.)
Dry open woods at all elevations in the
west and middle Gorge.
LATE MAY-EARLY JUNE: Stanley
Rock.
EARLY JULY: Grassy Knoll Trailhead.
MID JULY: South Prairie Road at MP
5-6.
Orchid Family.

64
Listera caurina   II
Western Twayblade   (4-12 in.)
Moist coniferous woods at middle to
high elevations in the west Gorge.
EARLY-MID JULY: Spring Camp Road.
Orchid Family.

65
Listera convallarioides   I
Broad-lip Twayblade   (4-8 in.)
Moist woods and streambanks at
middle to high elevations in the west
Gorge.
MID JULY: Herman Creek Trail at
MP 8.
Orchid Family.

66
Listera cordata   II
Heart-leaf Twayblade   (3-8 in.)
Deep mossy woods at all elevations in
the west Gorge.
APRIL: Ainsworth State Park.
MID JUNE: Herman Creek Trail at
MP 8.
EARLY JULY: near North Lake
Trailhead.
Orchid Family.

67
Spiranthes porrifolia   I
Western Ladies' Tresses   (6-24 in.)
Meadows, riverbanks, and transitory
streams at low to middle elevations
between Hood River and The Dalles
area.
MID JULY: Campbell Creek; Horsethief
Butte.
Orchid Family.

68
Spiranthes romanzoffiana   I
Hooded Ladies' Tresses   (8-24 in.)
Grassy bottomlands; wet meadows at
middle to high elevations in the west
half of the Gorge.
AUG: open woods on the north side of
Mirror Lake.
SEPT: wet meadows above 2,800 ft.
elev.
Orchid Family.

BEECH FAMILY

69
Castanopsis chrysophylla   II
(Chrysolepis chrysophylla)
Chinquapin   (3-20 ft.)
Generally found as a shrub on dry
ridges on the Oregon side between
Munra Point and Mitchell Point.
JUNE: Eagle Creek Overlook Picnic
Area.
EARLY AUG: slopes of Mt. Defiance;
Rainy Lake Road.
Beech Family.

70
Urtica dioica var. lyallii   V
Stinging Nettle   (2-8 ft.)
Shady moist areas at low elevations in the west Gorge.
MAY: Scenic Highway at Crown Point and Latourell Falls.
The *var. californica* is found at the east end of the Gorge.
Nettle Family.

71
Comandra umbellata   V
var. californica
Common Bastard   (4-12 in.)
Toad-flax
Dry open slopes or open woods at all elevations as far east as The Dalles.
MID MAY: Old US-30 west of Mosier; Memaloose Viewpoint; Rowland Lake.
EARLY JUNE: Hood River Mountain Meadow.
Sandalwood Family.

72
Comandra umbellata   I
var. pallida
Pale Bastard Toad-flax   (4-12 in.)
Dry grasslands at low elevations east of The Dalles.
LATE APRIL: DNR Falls.
Sandalwood Family.

73
Asarum caudatum   IV
Wild Ginger   (4-8 in.)
Moist woods at low to middle eleva-
tions in the west Gorge.
LATE APRIL-EARLY MAY: Latourell
Falls; Bridal Veil Creek; Ainsworth
State Park.
LATE JUNE: Pacific Crest Trail south of
Wahtum Lake.
Birthwort Family.

74
Eriogonum compositum   V
var. compositum
(white form)
Heart-leaf Buckwheat   (8-20 in.)
Dry open slopes, cliffs, talus slopes,
and rocky flats west of The Dalles.
LATE MAY-EARLY JUNE: Dog Creek
Falls.
MID JUNE: Rooster Rock; top of The
Dalles Mountain Road, where both
white and yellow forms are present.
EARLY JULY: Cape Horn.
Buckwheat Family.

75
Eriogonum compositum   IV
var. compositum
(yellow form)
Heart-leaf Buckwheat   (8-18 in.)
Dry open slopes, cliffs, talus slopes,
and rocky flats at all elevations east of
The Dalles.
MID-LATE MAY:rocky places along
SR-14 and I-84; Deschutes River
shore.
Buckwheat Family.

76
Eriogonum douglasii   II
var. tenue
Douglas' Buckwheat   (4-12 in.)
Dry open flats and slopes at all elevations in the east Gorge.
MID-LATE MAY: SR-14 at the rest area west of Lyle (MP 74); top of The Dalles Mountain Road; Haystack Butte Road.
Buckwheat Family.

77
Eriogonum elatum   III
Tall Buckwheat   (1.5-3.5 ft.)
Open flats and ridges at all elevations in the east Gorge, and at high elevations in the middle Gorge.
JULY: Tom McCall Nature Preserve; SR-14 east of Lyle Tunnel (MP 77-81); top of The Dalles Mountain Road; Monte Carlo.
Buckwheat Family.

78
Eriogonum strictum   IV
var. proliferum
Strict Buckwheat   (1-2 ft.)
Sandy or rocky soil at low elevations in the east Gorge.
MID-LATE AUG: Tom McCall Nature Preserve; Mayer State Park; SR-14 east of Lyle (MP 77-81); Horsethief Butte.
Buckwheat Family.

79

Eriogonum thymoides   I
Thyme Buckwheat   (6-10 in.)
Dry stony ground at high elevations in
the east Gorge.
JUNE: crest of the Columbia Hills west
of The Dalles Mountain Road.
Buckwheat Family.

80

Eriogonum umbellatum   II
var. umbellatum
Sulfur Buckwheat   (2-10 in.)
Rock crevices and talus slopes at either
low or high elevations in the middle
Gorge.
MID-LATE MAY: SR-14 at Tunnel #1.
LATE JUNE: Grassy Knoll Trailhead.
MID JULY: Monte Carlo; Indian Moun-
tain.
Buckwheat Family.

81

Eriogonum vimineum   II
Wire-stem Buckwheat   (2-10 in.)
Dry ground at low elevations in the
east Gorge.
JULY: SR-14 at MP 70.4 and at the
Wishram Historical Marker.
Buckwheat Family.

82
Polygonum amphibium   II
Water Smartweed   (6-12 in.)
Lakes and ponds—including vernal
ponds—throughout the Gorge, gener-
ally at low elevations.
AUG: pond south of I-84 near MP 75.
Buckwheat Family.

83
Polygonum bistortoides   II
Mountain Meadow   (1-2 ft.)
Knotweed
Wet meadows and rocky open slopes
at high elevations in the west Gorge.
LATE JUNE: Larch Mountain; Silver
Star Mountain.
MID-LATE JULY: wet meadows in the
west end of the Gorge.
Buckwheat Family.

84
Polygonum cascadense   I
Cascade Knotweed   (4-10 in.)
Dry open ground in the west Gorge.
LATE AUG-EARLY SEPT: Eagle Creek
Trail.
Buckwheat Family.

85
Polygonum coccineum   II
Pink Knotweed   (1-2 ft.)
Columbia River shores and bottom-
lands as far east as Horsethief Lake
State Park.
LATE AUG-EARLY SEPT: Dalton Point;
Home Valley Park; shore at Cape
Horn; Crates Point Wildlife Area.
Buckwheat Family.

86
Polygonum   II
hydropiperoides
Water Pepper   (1-2 ft.)
Wet bottomlands of the Columbia
River as far east as The Dalles area.
MID AUG: Mirror Lake; Franz Lake;
Home Valley Park.
Buckwheat Family.

87*
Polygonum lapathifolium   III
Willow Weed   (16-30 in.)
Columbia River shores and bottom-
lands, and other low elevation wet
areas.
MID SEPT: Dalton Point; Sand Island.
Buckwheat Family.

88
Polygonum majus ‖
Wiry Knotweed (6-12 in.)
Dry, gravelly or sandy ground at low elevations in the east Gorge.
SUMMER: dunes at The Dalles Airport; Horsethief Butte; Bobs Point; sandy areas along The Dalles Bridge Road (US-197).
Buckwheat Family.

89
Polygonum ‖
spergulariaeforme
Autumn Knotweed (2-6 in.)
Dry open ground and disturbed areas at low elevations in the west Gorge.
SEPT: Government Cove; SR-14 at Carson Depot Road.
Closely-related *Polygonum nuttallii* blooms in September along the Scenic Highway at Ainsworth State Park.
Buckwheat Family.

90
Rumex maritimus |
Golden Dock (6-24 in.)
Low elevation ponds and Columbia River shores as far east as The Dalles.
LATE JULY-EARLY AUG: Tom McCall Nature Preserve.
MID SEPT: Dalton Point.
Buckwheat Family.

91
Rumex occidentalis   II
var. procerus
Western Dock   (3-5 ft.)
Drying margins of vernal ponds,
meadows, and transitory streams
between Hood River and The Dalles.
EARLY MAY: Old Highway about 1.5
miles east of Major Creek.
EARLY JUNE: Horsethief Butte; mead-
ows on the Major Creek Plateau.
Buckwheat Family.

92
Rumex venosus   II
Veiny Dock   (6-18 in.)
Dunes along the Columbia River in the
east Gorge.
MID APRIL: Crates Point Dunes; Bobs
Point.
Buckwheat Family.

GOOSEFOOT FAMILY

93 *
Salsola kali   II
Tumbleweed   (1-3 ft.)
Roadsides, railroad embankments, and
other disturbed ground, typically in the
east Gorge.
LATE AUG-EARLY SEPT: I-84 between
Hood River and Biggs Junction; SR-14
near Lyle Tunnel; The Dalles Dam
Visitor Center.
Goosefoot Family.

94
Abronia mellifera   II
White Sand-verbena   (6-16 in.)
Dunes and sandy ground east of The Dalles.
LATE JUNE-EARLY JULY: dunes at The Dalles Airport; I-84 (eastbound) near the turnout at MP 92.3.
Four-o'clock Family.

95
Claytonia lanceolata   III
Lance-leaf Spring Beauty   (4-7 in.)
Open or lightly wooded areas at middle to high elevations in the middle and east Gorge.
EARLY-MID APRIL: tops of Dry Creek Road, Carroll Road, and Vensel Road.
MID APRIL: Hood River Mountain Meadow.
MID-LATE MAY: Monte Carlo.
Purslane Family.

96
Lewisia columbiana   II
var. columbiana
Columbia Lewisia   (4-12 in.)
Open rocky areas in the west Gorge, generally at middle to high elevations.
MID JUNE: Oneonta Gorge; Hamilton Mountain Trail.
LATE JUNE-EARLY JULY: Silver Star Mountain.
Purslane Family.

97
Lewisia rediviva II
Bitter Root (1-2 in.)
Dry gravelly soil in open areas in the
east Gorge.
EARLY MAY: Old Highway near
Catherine Creek; SR-14 near Catherine
Creek (MP 72.2); near Memaloose
Viewpoint.
EARLY JUNE: Columbia Hills ridge
west of The Dalles Mountain Road and
also east of Haystack Butte.
Purslane Family.

98
Montia Chamissoi I
Water Montia (4-7 in.)
Wet areas at low to middle elevations
in the east Gorge.
LATE APRIL-EARLY MAY: Horsethief
Butte.
LATE MAY-EARLY JUNE: slow streams
on the Major Creek Plateau; meadow
at DNR Forest.
Purslane Family.

99
Montia cordifolia I
Broad-leaf Montia (6-12 in.)
Springs at higher elevations near the
heads of Eagle Creek and Herman
Creek, and near Three-Corner Rock.
JULY: springs along the primitive road
(gated) north of Wahtum Lake Forest
Camp.
Purslane Family.

100
Montia diffusa   I
Branching Montia   (4-8 in.)
LATE MAY-EARLY JUNE: open rocky,
often disturbed areas in the west
Gorge.
Purslane Family.

101
Montia exigua   I
(Montia spathulata)
Pale Montia   (1-4 in.)
Open ground in the east Gorge.
LATE MAR-EARLY APRIL: Horsethief
Butte.
MID APRIL: US-30 at Crates Point;
cliffs along I-84 (eastbound) at Mayer
State Park.
Purslane Family.

102
Montia linearis   IV
Line-leaf Montia   (2-8 in.)
Moist, open or lightly wooded slopes
throughout the Gorge. Especially
common between Hood River and The
Dalles.
EARLY APRIL: Memaloose Rest Area;
Tom McCall Nature Preserve; Stanley
Rock.
Purslane Family.

103
Montia parvifolia V
Little-leaf Montia (4-10 in.)
Moist cliffs and mossy banks at all
elevations in the west Gorge.
MID-LATE MAY: Crown Point; La-
tourell Falls; Beacon Rock; Dog Creek
Falls.
Purslane Family.

104
Montia perfoliata V
Miner's-lettuce (2-12 in.)
Moist ground in lightly to densely
shaded woods throughout the Gorge,
generally at low elevations between
Hood River and The Dalles.
LATE MAR: Memaloose Rest Area; SR-
14 east of Lyle Tunnel (MP 77-81).
EARLY APRIL: Stanley Rock.
LATE APRIL-EARLY MAY: near Oneon-
ta Bridge; Hood River Mountain
Meadow.
Purslane Family.

105
Montia sibirica V
Candy Flower (6-14 in.)
Moist woods in the west Gorge.
MID APRIL: Ainsworth State Park;
Lower Tanner Creek Road.
LATE MAY: Latourell Falls.
MID JUNE: Crown Point.
Purslane Family.

106
Arenaria capillaris I
Mountain Sandwort (4-8 in.)
Open gravelly soil at high elevations in
the west Gorge.
MID JUNE: top of Big Huckleberry
Mountain.
LATE JUNE-EARLY JULY: Indian Mountain.
Pink Family.

107
Arenaria franklinii I
var. franklinii
Franklin's Sandwort (2-6 in.)
Open gravelly soil at high elevations in
the east Gorge.
LATE MAY-EARLY JUNE: Columbia
Hills ridge west of The Dalles Mountain
Road and east of Haystack Butte.
Pink Family.

108
Arenaria macrophylla V
Big-leaf Sandwort (3-8 in.)
Woods in the west and middle Gorge.
MID-LATE APRIL: Beacon Rock.
MID JUNE: top of Larch Mountain.
Pink Family.

109
Arenaria rubella I
Reddish Sandwort (4-6 in.)
Open gravelly slopes, rock crevices,
and gravel bars along streams in the
west Gorge.
EARLY-MID JUNE: Beacon Rock Trail;
near the top of Dog Mountain.
Pink Family.

110
Arenaria stricta I
Slender Sandwort (4-8 in.)
LATE MAY-EARLY JUNE: grassy open
places in the west Gorge.
Pink Family.

111
Cerastium arvense V
Field Chickweed (4-10 in.)
Open, lightly wooded and rocky slopes
at all elevations in the west Gorge.
LATE APRIL-EARLY MAY: Dog Creek
Falls; SR-14 near Tunnel #1.
MID-LATE MAY: Scenic Highway at
Crown Point, Shepperd's Dell, and
Oneonta Gorge.
Pink Family.

112*
Dianthus armeria II
Grass Pink (1-2 ft.)
Disturbed areas in the west Gorge, generally at low elevations.
LATE JUNE-EARLY JULY: Grant Lake.
MID JULY: SR-14 at the Clark-Skamania County Line.
LATE JULY: SR-14 at Cape Horn.
Pink Family.

113*
Lychnis alba III
White Campion (1.5-3 ft.)
Roadsides and other disturbed areas in the west Gorge.
LATE JUNE-EARLY JULY: SR-14 at the Clark-Skamania County Line.
LATE JULY-EARLY AUG: Corbett Hill Road.
Pink Family.

114*
Saponaria officinalis III
Soapwort (1-3 ft.)
Gravelly streambanks, ditches, and roadsides throughout the Gorge.
LATE JUNE-EARLY JULY: SR-14 east of Horsethief Butte.
LATE JULY-EARLY AUG: I-84 at Cascade Locks; Scenic Highway east of Multnomah Falls.
Pink Family.

115
Silene antirrhina I
Sleepy Catchfly (8-16 in.)
Open, grassy or mossy slopes in the
west and middle Gorge.
EARLY JUNE: Grassy Knoll Trailhead.
Pink Family.

116
Silene douglasii III
Douglas' Campion (4-16 in.)
Open or lightly wooded slopes and
cliffs in the west Gorge.
MID JUNE: near Crown Point; Shep-
perd's Dell; Pillars of Hercules.
LATE JUNE-EARLY JULY: Cape Horn.
LATE JULY: top of Indian Mountain.
Pink Family.

117
Silene menziesii I
var. viscosa
Menzies' Campion (4-12 in.)
Low elevation streambanks and moist
meadows between the White Salmon
River and Horsethief Butte.
LATE MAY-EARLY JUNE: Klickitat
Fisherman's Park; Horsethief Butte.
Pink Family.

118
Silene oregana II
Oregon Campion (1-2 ft.)
Open or lightly wooded ridges at high
elevations in the west Gorge.
EARLY JUNE: Grassy Knoll Trailhead.
EARLY JULY: Monte Carlo.
Pink Family.

119
Stellaria longifolia II
Long-leaf Starwort (6-24 in.)
Wet meadows, ditches, and other
poorly drained areas in the west and
middle Gorge.
EARLY JULY: Brower Road; meadows
on the Major Creek Plateau.
Pink Family.

WATER-LILY FAMILY

120
Nuphar polysepalum II
Yellow Water-lily
Ponds at all elevations in the west and
middle Gorge.
MID JUNE: Tom McCall Nature Pre-
serve; pond just northwest of Grant
Lake.
LATE JULY-EARLY AUG: Beacon Rock
Pond.
Water-lily Family.

121*
Nymphaea odorata   I
American Water-lily
Native to eastern North America.
Found in low elevation lakes and
ponds in the west and middle Gorge.
LATE JULY-EARLY AUG: Ice House
Lake; Tom McCall Nature Preserve.
Water-lily Family.

BUTTERCUP FAMILY

122
Actaea rubra   IV
Baneberry   (1.5-3 ft.)
Moist woods at all elevations in the
west Gorge.
EARLY MAY: Scenic Highway at Bridal
Veil, Horsetail Falls, and Ainsworth
State Park.
LATE MAY-EARLY JUNE: Larch Moun-
tain Highway beyond MP 3.4.
Buttercup Family.

123
Anemone deltoidea   IV
Columbia Wind Flower,   (6-12 in.)
Western White Anemone
Moist woods at all elevations in the
west Gorge.
LATE MAY: Scenic Highway at La-
tourell Falls, Shepperd's Dell, and
Bridal Veil Creek.
EARLY JULY: Spring Camp Road.
Buttercup Family.

124
Anemone lyallii   I
Lyall's Anemone   (3-6 in.)
Moist woods, normally at low elevations in the west Gorge.
LATE MAR-EARLY APRIL: Scenic Highway at Bridal Veil Creek and just east of Bridal Veil.
Buttercup Family.

125
Anemone multifida   I
var. multifida
Cliff Anemone   (6-16 in.)
Open ridges at high elevations.
EARLY JUNE: the heads of Eagle Creek and Herman Creek.
Buttercup Family.

126
Anemone oregana   III
Oregon Anemone   (4-12 in.)
Open woods at middle to high elevations in the west Gorge, particularly in the heights overlooking Bonneville Dam.
EARLY MAY: Hamilton Mountain Trail.
Buttercup Family.

127
Aquilegia formosa　　　IV
Red Columbine　　　(1-3 ft.)
Open woods at all elevations in the
west and middle Gorge.
EARLY JUNE: Scenic Highway at
Shepperd's Dell and Bridal Veil; Beacon
Rock Trail.
EARLY JULY: Rainy Lake and the top
of Larch Mountain.
Buttercup Family.

128
Caltha biflora　　　III
Marsh-marigold　　　(6-14 in.)
Wet meadows, lake margins, and
springs at middle to high elevations in
the west Gorge.
LATE JUNE: North Lake Trailhead;
Rainy Lake.
Buttercup Family.

129
Cimicifuga elata　　　II
Tall Bugbane　　　(3-6 ft.)
Low elevation woods in the west
Gorge.
MID-LATE JUNE: Scenic Highway near
Shepperd's Dell, near Angels Rest
Trailhead, and at Ainsworth State
Park; woods at Cape Horn.
Buttercup Family.

130
Cimicifuga laciniata   III
Cut-leaf Bugbane   (2.5-5 ft.)
Damp woods and marshy areas at higher elevations in the west Gorge, mostly on the Oregon side.
LATE AUG: Wahtum Lake; Indian Springs.
Buttercup Family.

131
Clematis ligusticifolia   III
Western Clematis   (vine)
Low elevation areas along the Columbia River throughout the Gorge, well above the high-water mark.
LATE JUNE-EARLY JULY: US-30 just west of The Dalles; SR-14 east of Horsethief Butte.
LATE JULY-EARLY AUG: I-84 at the Corbett Exit (#22).
Buttercup Family.

132
Coptis laciniata   I
Cut-leaf Gold Threads   (3-7 in.)
(shown in fruit)
Moist woods at middle elevations in the west end of the Gorge.
MAR: McCloskey Creek Road near the bridge. Fruit appears in mid-late April.
Buttercup Family.

133
Delphinium burkei I
Meadow Larkspur (10-20 in.)
Areas that are vernally wet, drying by
summer, between Bingen and Horse-
thief Lake.
EARLY JUNE: near Catherine Creek;
Tom McCall Nature Preserve; Horse-
thief Lake State Park.
Buttercup Family.

134
Delphinium menziesii IV
var. pyramidale
Cliff Larkspur (8-18 in.)
Shaded cliffs and rocky banks at all
elevations in the west Gorge.
EARLY MAY: base of Rooster Rock.
MID MAY: Scenic Highway at Multno-
mah Falls and Oneonta Gorge.
Buttercup Family.

135
Delphinium nuttallianum V
Upland Larkspur (4-16 in.)
Dry grasslands and open woods from
Dog Mountain to the east end of the
Gorge.
MID-LATE APRIL: Major Creek Road;
Tom McCall Nature Preserve; Mayer
State Park; Hood River Mountain
Meadow.
Buttercup Family.

136
Delphinium nuttallii   II
Nuttall's Larkspur   (1-2 ft.)
Open or partly shaded, moist grassy
slopes and meadows as far east as the
Wind River.
MID JUNE: SR-14 at the Clark-Skama-
nia County Line and at Signal Rock.
LATE JUNE: Crown Point; Cape Horn.
Buttercup Family.

137
Delphinium trollifolium   IV
Poison Larkspur   (2-5 ft.)
Moist shady woods in the west Gorge,
commonly at low elevations west of
Bonneville Dam.
MID MAY: Scenic Highway from
Crown Point to Ainsworth State Park;
SR-14 west of Cape Horn.
Buttercup Family.

138
Isopyrum hallii   II
Hall's Isopyrum   (1-3 ft.)
Moist woods at middle to high eleva-
tions as far east as Mt. Defiance. Thus
far found only on the Oregon side.
MID-LATE JUNE: ditch along Palmer
Mill Road above the Brower Road
junction.
LATE JULY: Wahtum Lake.
EARLY AUG: Indian Springs Road 0.5
mile south of Wahtum Lake Forest
Camp.
Buttercup Family.

139
Ranunculus alismaefolius III
var. alismaefolius
Plantain-leaf Buttercup (1-2 ft.)
Vernally wet meadows and ponds at
all elevations between the White
Salmon River and The Dalles.
LATE APRIL: near Lyle Cemetery;
Stevens Pond.
MID-LATE MAY: wet meadows on the
Major Creek Plateau; pond at Wasco
Butte.
Buttercup Family.

140
Ranunculus aquatilis II
var. capillaceus
White Water Buttercup
Ponds and slow streams in the middle
and east Gorge, especially between
Hood River and The Dalles area.
LATE MAY: wet meadows on the
Major Creek Plateau; Stevens Pond.
LATE MAY-EARLY JUNE: Tom McCall
Nature Preserve; Horsethief Lake State
Park.
Buttercup Family.

141
Ranunculus cymbalaria II
Shore Buttercup (2-5 in.)
Muddy shores of the Columbia River,
especially in the east Gorge.
AUG: west bank of the Deschutes
River.
LATE AUG-EARLY SEPT: Oregon shore
west of The Dalles Bridge.
Buttercup Family.

142
Ranunculus flammula   III
Small Creeping Buttercup   (4-10 in.)
Muddy areas of ponds and Columbia
River bottomlands.
LATE JULY: Tom McCall Nature Pre-
serve; Stevens Pond.
LATE AUG: Mirror Lake.
Buttercup Family.

143
Ranunculus glaberrimus   I
Sagebrush Buttercup   (3-6 in.)
Open or lightly wooded areas at all
elevations east of Lyle.
MID MAR: hillside along US-30 at
Crates Point.
MID-LATE MAR: Klickitat Fisherman's
Park.
Buttercup Family.

144
Ranunculus macounii   II
Macoun's Buttercup   (4-18 in.)
Damp ground near the high-water
mark along Columbia River shores and
bottomlands in the west Gorge, espe-
cially west of Bonneville Dam.
LATE APRIL: Dalton Point; Young
Creek in Rooster Rock State Park.
OCT: Dalton Point; Young Creek in
Rooster Rock State Park.
Buttercup Family.

145
Ranunculus occidentalis V
var. occidentalis
Western Buttercup   (8-18 in.)
Open or lightly wooded ground at all
elevations from Cascade Locks to The
Dalles, often under oaks. Common
between Hood River and The Dalles.
EARLY APRIL: Memaloose Rest Area;
backroads out of Mosier; Major Creek
Road; US-30 west of The Dalles near
MP 15.
Buttercup Family.

146
Ranunculus orthorhynchus II
var. orthorhynchus
Western Swamp Buttercup   (1-2 ft.)
Wet meadows, ditches, and slow
streams in the middle Gorge.
MID MAY: Old Highway near MP 4.
LATE MAY-EARLY JUNE: meadows on
the Major Creek Plateau.
Buttercup Family.

147
Ranunculus reconditus   I
The Dalles Mountain   (2-5 in.)
Buttercup
Open grasslands on top of the Colum-
bia Hills in Washington and similar
areas near the heads of Mill Creek and
Mosier Creek in Oregon.
EARLY-MID MAR: top of The Dalles
Mountain Road.
Buttercup Family.

148
Ranunculus sceleratus   III
var. multifidus
Celery-leaf Buttercup   (6-16 in.)
Columbia River shores and bottom-
lands and other muddy-wet places
throughout the Gorge.
EARLY JUNE: Mirror Lake; Grant Lake;
ditch near Horsethief Butte.
EARLY SEPT: Dalton Point.
Buttercup Family.

149
Ranunculus uncinatus   IV
var. parviflorus
Little Buttercup   (1-2 ft.)
Coniferous woods as far east as The
Dalles.
MID-LATE MAY: Scenic Highway at
Latourell Falls, Bridal Veil, and Multno-
mah Falls.
LATE JUNE: Spring Camp Road.
Buttercup Family.

150
Thalictrum occidentale   IV
Western Meadow Rue   (1.5-3 ft.)
Moist coniferous woods at all eleva-
tions in the west Gorge.
LATE APRIL: Scenic Highway at Bridal
Veil, Mt. Hood National Forest bound-
ary, and Ainsworth State Park.
Buttercup Family.

151
**Trautvetteria caroliniensis   III
var. occidentalis**
False Bugbane   (1.5-3 ft.)
Moist woods at middle to high eleva-
tions in the west Gorge.
EARLY JULY: Palmer Mill Road; Rainy
Lake.
LATE JULY: Indian Springs Road 0.5
mile south of Wahtum Lake Forest
Camp.
Buttercup Family.

BARBERRY FAMILY

152
Achlys triphylla   IV
Vanilla Leaf   (8-16 in.)
Coniferous woods at all elevations in
the west Gorge.
EARLY MAY: Scenic Highway at Bridal
Veil Creek and Bridal Veil; Smith-Cripe
Road east of Cape Horn.
Barberry Family.

153
Berberis aquifolium   V
Shining Oregon Grape   (1-4 ft.)
Open or lightly wooded areas in low to
middle elevations throughout the
Gorge.
MID APRIL: Beacon Rock; Memaloose
Rest Area; Stanley Rock; Rowland
Lake.
Barberry Family.

154
Berberis nervosa V
Cascade Oregon Grape (6-24 in.)
Coniferous woods at all elevations in
the west Gorge.
MID APRIL: Crown Point vicinity.
LATE APRIL: Eagle Creek Forest Camp.
EARLY JUNE: Spring Camp Road.
LATE JUNE: top of Larch Mountain.
Barberry Family.

155
Berberis repens I
Creeping Oregon Grape (3-6 in.)
Open grassy, generally north-facing
slopes east of The Dalles.
EARLY-MID APRIL: BLM Parcel at
Celilo.
Barberry Family.

156
Vancouveria hexandra V
Inside-out Flower (8-16 in.)
Moist shady woods at all elevations in
the west Gorge.
LATE MAY-EARLY JUNE: Scenic High-
way at Latourell Falls, Bridal Veil Creek,
and Ainsworth State Park.
MID JUNE: Brower Road.
LATE JUNE-EARLY JULY: Spring Camp
Road.
Barberry Family.

157
Eschscholzia californica   IV
California Poppy   (6-16 in.)
Open areas at low elevations as far
east as The Dalles area. Most common
on the Washington side between Dog
Mountain and Lyle.
EARLY MAY: SR-14 at Tunnel #4 and
at Lyle.
LATE MAY: I-84 (eastbound) at Hood
River.
Poppy Family.

158
Meconella oregana   II
White Meconella   (1-4 in.)
Open areas at low elevations between
Hood River and Lyle.
EARLY APRIL: Stanley Rock; Mema-
loose Viewpoint; Tom McCall Nature
Preserve.
Poppy Family.

BLEEDING HEART FAMILY

159
Corydalis aquae-gelidae   I
Cold-water Corydalis   (1-3 ft.)
In and beside small perennial streams
in wooded areas of the west Gorge.
Blooms from mid-May to mid-July,
depending on elevation.
LATE MAY: Tanner Creek Trail in the
Columbia Wilderness.
Bleeding Heart Family.

160
Corydalis scouleri　　　　III
Western Corydalis　　　　(2-4 ft.)
Moist soil at low elevations as far east
as Cascade Locks.
LATE APRIL-EARLY MAY: Lower
Tanner Creek Road; Scenic Highway at
Latourell Falls, Bridal Veil Creek, and
the Mt. Hood National Forest
boundary.
Bleeding Heart Family.

161
Dicentra cucullaria　　　　II
Dutchman's Breeches　　　(6-18 in.)
Moist soil, mostly at low elevations as
far east as Lyle.
LATE MAR-EARLY APRIL: Starvation
Creek State Park; Memaloose Rest
Area.
LATE APRIL: Multnomah Falls.
Bleeding Heart Family.

162
Dicentra formosa　　　　V
Bleeding Heart　　　　(6-18 in.)
Moist woods at all elevations in the
west Gorge.
MID APRIL: Latourell Falls; Ainsworth
State Park.
MID JUNE: top of Larch Mountain.
Bleeding Heart Family.

SECTION TWO

163
Arabis furcata   I
Cascade Rock Cress   (5-12 in.)
Rocky banks and talus slopes at low
elevations on the Oregon side between
Shellrock Mountain and Viento State
Park.
LATE APRIL-EARLY MAY: talus slopes
west of Starvation Creek State Park.
Mustard Family.

164
Arabis hirsuta   III
var. eschscholtziana
Hairy Rock Cress   (6-24 in.)
Moist soil in forests and forest openings
at all elevations in the west Gorge.
EARLY MAY: Crown Point; Beacon
Rock; cliffs along Corbett Hill Road.
Mustard Family.

165
Arabis hirsuta var. glabrata   I
Hairy Rock Cress   (6-18 in.)
EARLY APRIL: rocky face at the old
navigation lock at Bonneville Dam.
Mustard Family.

166
Arabis microphylla ll
Little-leaf Rock Cress (4-10 in.)
Open rocky outcrops and cliffs at low
elevations as far east as Horsethief
Butte.
MID MAR: cliffs at the mouth of
Fifteen-Mile Creek near The Dalles
Dam Visitor Center.
LATE MAR-EARLY APRIL: near the
gate of Beacon Rock Trail; Horsethief
Butte; I-84 (eastbound) west of Mema-
loose Rest Area.
Mustard Family.

167
Arabis sparsiflora l
var. atrorubens
Sickle-pod Rock Cress (6-24 in.)
Open, rocky or gravelly places at high
elevations on the Washington side
between Grassy Knoll and the Colum-
bia Hills.
MID APRIL: west side of Road 68
below Grassy Knoll.
LATE MAY-EARLY JUNE: Monte Carlo.
Mustard Family.

168
Barbarea orthoceras lll
American Winter Cress (1-2 ft.)
Moist woods and forest openings at
low to middle elevations in the west
and middle Gorge.
EARLY MAY: Lower Tanner Creek
Road.
Mustard Family.

169
Cardamine angulata   I
Angle-leaf Bitter Cress   (1-2 ft.)
Moist woods at low elevations in the
west end of the Gorge.
LATE MAY: Scenic Highway near
Crown Point and Latourell Falls.
Mustard Family.

170
Cardamine breweri   I
var. orbicularis
Brewer's Bitter Cress   (6-12 in.)
Low elevation streams in the west
Gorge.
LATE APRIL-EARLY MAY: streams
along Belle Center Road and SR-14 at
MP 35.3.
Mustard Family.

171
Cardamine cordifolia   I
Large Mountain Bitter Cress   (1-2 ft.)
MID-LATE MAY: thus far found in the
Gorge only at the bottom of the Little
White Salmon River canyon, about a
mile above the confluence with the
Columbia River.
Mustard Family.

172
Cardamine integrifolia   II
Coast Toothwort   (6-14 in.)
Moist coniferous woods as far east as
Wind Mountain.
EARLY APRIL: Hamilton Mountain
Trail; McCloskey Creek Road near the
bridge.
Mustard Family.

173
Cardamine occidentalis   I
Western Bitter Cress   (4-14 in.)
Damp meadows and streamsides in
the west and middle Gorge.
MID MAY: meadows on the Major
Creek Plateau.
EARLY JUNE: higher elevation marshes
in the west Gorge.
Mustard Family.

174
Cardamine oligosperma   IV
Little Western Bitter Cress   (4-14 in.)
Moist open areas in the west and
middle Gorge, generally at low eleva-
tions.
LATE MAR-EARLY APRIL: Ainsworth
State Park; Stanley Rock; Dog Creek
Falls.
Mustard Family.

175
Cardamine pensylvanica III
Pennsylvania Bitter Cress (6-18 in.)
Muddy banks of low elevation streams
in the west Gorge, especially Columbia
River shores.
MAY-SEPT: Dalton Point; mouth of
Bridal Veil Creek.
Mustard Family.

176
Cardamine pulcherrima V
var. pulcherrima
Oaks Toothwort (4-10 in.)
Generally at low elevations from the
Wind River to The Dalles area, com-
monly in the partial shade of oaks.
LATE MAR-EARLY APRIL: Stanley
Rock; Tom McCall Nature Preserve;
Old US-30 west of Mosier; woods
near the Klickitat Fish Ladder.
Mustard Family.

177
Cardamine pulcherrima V
var. tenella
Slender Toothwort (4-10 in.)
Woods as far east as Cascade Locks,
generally at low elevations.
LATE MAR-EARLY APRIL: Scenic
Highway at Ainsworth State Park;
woods along SR-14 near MP 20.
MID APRIL: Beacon Rock.
Mustard Family.

178
Descurainia richardsonii   III
Tansy Mustard   (6-18 in.)
Open, rocky or sandy areas, generally
at low elevations east from Bingen.
LATE MAR-EARLY APRIL: SR-14 east
of Lyle Tunnel (MP 77-81) and east of
Horsethief Butte; BLM Parcel at Celilo.
Mustard Family.

179
Draba douglasii   I
Douglas' Draba   (1-4 in.)
MID-LATE APRIL: open gravelly flats
on top of the Columbia Hills west of
The Dalles Mountain Road.
Mustard Family.

180
Draba verna   V
var. boerhaavii
Spring Whitlow-grass   (2-6 in.)
Open grassy areas throughout the
Gorge, most common at low elevations
from Dog Mountain to The Dalles.
LATE MAR: Stanley Rock; Memaloose
Viewpoint; SR-14 east of Lyle Tunnel
(MP 77-81).
Mustard Family.

181
Erysimum asperum V
Rough Wallflower   (6-30 in.)
Open or lightly wooded slopes as far
east as The Dalles area, chiefly at low
elevations.
LATE APRIL: the Old Highway.
MID MAY: SR-14 along the base of
Dog Mountain.
EARLY JUNE: Cape Horn; Scenic
Highway near Crown Point.
Mustard Family.

182
Erysimum occidentale II
Pale Wallflower   (6-16 in.)
Dry sandy ground and dunes at low
elevations from Crates Point to the
east end of the Gorge.
EARLY APRIL: Horsethief Butte; Avery
Gravel Pit; Bobs Point; BLM Parcel at
Celilo.
Mustard Family.

183
Idahoa scapigera IV
Scale Pod (shown in fruit)   (2-4 in.)
Open ground between Bingen and
The Dalles.
MID MAR: top of The Dalles Mountain
Road; Old Highway; SR-14 east of
Lyle Tunnel (MP 77-81).
The distinctive pods remain till mid-
April.
Mustard Family.

184
Lesquerella douglasii   I
Bladder Pod   (4-8 in.)
Sandy or gravelly open places near the shores of the Columbia River in the east Gorge.
MID APRIL: Avery Gravel Pit; east end of Miller Island.
Mustard Family.

185*
Lunaria annua   II
Honesty   (1.5-3 ft.)
Roadsides and other disturbed areas in the west Gorge.
MID-LATE APRIL: Scenic Highway between Crown Point and Bridal Veil. Somewhat similar *Hesperis matronalis* is also found along the Scenic Highway, blooming around the first of June.
Mustard Family.

186
Phoenicaulis   II
cheiranthoides
Dagger Pod   (4-10 in.)
Open, gravelly or rocky ground, particularly at middle to high elevations east of The Dalles.
APRIL: The Dalles Mountain Road, midway and at the top.
Mustard Family.

187
Rorippa columbiae   I
Columbia Yellow Cress   (1-5 in.)
Gravelly shores of the Columbia River
downstream from Bonneville Dam,
growing below the high-water mark.
LATE AUG-EARLY SEPT: Columbia
River shores at the Pierce National
Wildlife Refuge, Pierce Island, Ives
Island, and the west end of Hamilton
Island.
Mustard Family.

188
Rorippa curvisiliqua   III
var. curvisiliqua
Western Yellow Cress   (4-16 in.)
Columbia River shores; ponds and
ditches throughout the Gorge.
SEPT: Columbia River shore at Dalton
Point; mouth of Bridal Veil Creek;
Beacon Rock shore.
Mustard Family.

189
Rorippa islandica   II
Marsh Yellow Cress   (6-18 in.)
Columbia River shores and bottom-
lands.
SEPT: Mirror Lake; Columbia River
shore west of the Corbett boat ramp
and near the mouth of Bridal Veil
Creek.
Mustard Family.

190*
Rorippa nasturtium-aquaticum   III

Water Cress   (1-2 ft.)
Ponds, ditches, and slow streams as far
east as The Dalles.
JUNE: ditches along SR-14 at Home
Valley, at the junction with Willard
Road (MP 56.3), and near Tunnel #4;
Old Highway at Rowland Lake.
Mustard Family.

191
Schoenocrambe linifolia   II

Plains Mustard   (8-18 in.)
Dry cliffs and rocky banks between
Hood River and The Dalles.
LATE APRIL-EARLY MAY: cliffs and
talus slopes between Bingen and Locke
Lake; cliffs along I-84 at Mayer State
Park.
Mustard Family.

192
Thelypodium laciniatum   III
var. laciniatum

Thick-leaf Thelypody   (1-6 ft.)
Basalt cliffs east of Mosier, generally at
low elevations.
LATE APRIL: SR-14 east of Lyle Tunnel
at MP 77.3; Horsethief Butte; Mayer
State Park.
Mustard Family.

193
Thlaspi fendleri   II
var. glaucum
Rock Penny Cress   (4-8 in.)
Open slopes at high elevations in the
west Gorge, mostly on the Washington
side.
MID MAY: near the top of Dog Moun-
tain.
EARLY JUNE: Big Huckleberry Moun-
tain; Indian Mountain; Silver Star
Mountain.
Mustard Family.

194
Thysanocarpus curvipes   V
Fringe Pod (shown in fruit)   (6-16 in.)
Open slopes at all elevations in the
east half of the Gorge.
LATE MAR: SR-14 east of Lyle Tunnel
(MP 77-81); Memaloose Viewpoint.
Mustard Family.

CAPER FAMILY

195
Cleome lutea   II
Yellow Bee Plant   (2-6 ft.)
Sandy areas at low to middle elevations
in the east Gorge.
SUMMER: The Dalles Bridge Ponds.
Caper Family.

196
Drosera rotundifolia II
Round-leaf Sundew (2-6 in.)
AUG: boggy meadows at higher
elevations in the west Gorge.
Closely-related *Drosera anglica* is
found at Deer Meadow on the Wash-
ington side.
Sundew Family.

197
Sedum leibergii III
Leiberg's Stonecrop (3-6 in.)
Cliffs and mossy rocks in the east
Gorge.
LATE MAY-EARLY JUNE: Memaloose
Viewpoint and cliffs along US-30 from
there to Crates Point.
Stonecrop Family.

198
Sedum oreganum III
Oregon Stonecrop (3-6 in.)
Cliffs, rocky outcrops, and talus slopes
at all elevations in the west Gorge.
LATE JULY: Lower Tanner Creek Road;
SR-14 east of Cape Horn; Scenic
Highway east of Crown Point.
Stonecrop Family.

199
Sedum spathulifolium V
Broad-leaf Stonecrop (3-8 in.)
Cliffs, talus slopes, and gravelly
benches as far east as Mosier.
MID-LATE MAY: SR-14 at Tunnel #1
and at Dog Creek Falls; I-84 (east-
bound) east of Starvation Creek State
Park; Rock Creek Road.
EARLY JUNE: Scenic Highway.
Stonecrop Family.

200
Sedum stenopetalum II
Worm-leaf Stonecrop (4-10 in.)
Rocky outcrops and gravelly benches
between Bonneville Dam and The
Dalles.
LATE JUNE: Bear Creek Road; Grassy
Knoll Trailhead; Monte Carlo.
Stonecrop Family.

SAXIFRAGE FAMILY

201
Bolandra oregana II
Oregon Bolandra (8-12 in.)
Waterfalls and moist cliffs, mostly at
low elevations in the west Gorge.
EARLY-MID JUNE: Lower Tanner
Creek Road; Multnomah Falls.
Saxifrage Family.

202
Boykinia elata I
Coast Boykinia (10-20 in.)
Streambanks in the west end of the
Gorge.
LATE JUNE: West Fork Washougal
River Bridge.
MID-LATE JULY: streams crossing the
Larch Mountain Highway at MP 8.9.
Saxifrage Family.

203
Heuchera chlorantha II
Meadow Alumroot (2-4 ft.)
Wet meadows, ditches, and moist
gravelly areas, mostly on the Major
Creek Plateau.
EARLY JUNE: meadows and ditches at
Laws Corner.
Saxifrage Family.

204
Heuchera grossulariifolia II
var. tenuifolia
Currant-leaf Alumroot (6-24 in.)
Shady cliffs and talus slopes from
Hood River to The Dalles area, gener-
ally on the Oregon side.
MID MAY: cliffs along Old US-30
between Hood River and Mosier.
Saxifrage Family.

205
Heuchera micrantha V
Small-flowered Alumroot (1-2 ft.)
Cliffs, rock crevices, and talus slopes at
all elevations in the west Gorge.
LATE MAY-EARLY JUNE: Scenic High-
way from Crown Point to Ainsworth
State Park; Beacon Rock.
Saxifrage Family.

206
Lithophragma bulbifera IV
Bulblet Prairie Star (4-8 in.)
Moist open slopes in the east Gorge.
LATE MAR-EARLY APRIL: SR-14 east
of Lyle Tunnel (MP 77-81); Horsethief
Lake State Park; Spearfish Lake Park;
near dunes at The Dalles Airport.
MID APRIL: top of The Dalles Moun-
tain Road.
Saxifrage Family.

207
Lithophragma glabra V
Smooth Prairie Star (4-8 in.)
Open areas and lightly wooded slopes
between Rooster Rock and Horsethief
Lake State Park.
MID-LATE MAR: grassy slopes along
SR-14 at MP 70.4 and east of Lyle
Tunnel at MP 77-81.
EARLY-MID APRIL: cliffs along US-30
at Rowena Crest (MP 7).
Saxifrage Family.

208
Lithophragma parviflora V
Small-flowered Prairie Star   (4-12 in.)
Rocky banks, grassy slopes, and open
woods throughout the Gorge, typically
at low elevations.
MID APRIL: I-84 (eastbound) east of
Starvation Creek State Park; SR-14
east of Lyle Tunnel (MP 77-81);
Chenoweth Road.
Saxifrage Family.

209
Mitella breweri II
Feathery Mitrewort   (6-10 in.)
Moist woods at higher elevations in
the west Gorge.
LATE JUNE-EARLY JULY: woods near
Chinidere Mountain; north of the top
of Silver Star Mountain; near the head
of Herman Creek.
Saxifrage Family.

210
Mitella caulescens III
Leafy-stem Mitrewort   (6-12 in.)
Damp shady ground in the west Gorge.
EARLY-MID MAY: Scenic Highway at
Bridal Veil Creek; Gorge Trail at Moffett
Creek.
LATE MAY: Haines Road at Latourell
Creek.
Saxifrage Family.

211
Mitella diversifolia II
Angle-leaf Mitrewort (6-18 in.)
(shown in fruit)
Moist woods and meadow edges at
middle elevations in the middle Gorge.
MID-LATE MAY: open woods on the
Major Creek Plateau; Wygant Trail.
Saxifrage Family.

212
Mitella pentandra II
Alpine Mitrewort (8-12 in.)
Moist woods in the west Gorge,
especially near streambanks, springs,
and bogs at elevations over 3,000 ft.
LATE JUNE-EARLY JULY: North Lake
Trailhead; Rainy Lake.
Saxifrage Family.

213
Mitella trifida II
Three-tooth Mitrewort (6-16 in.)
Open forests at middle to high eleva-
tions in the west Gorge.
EARLY JULY: near the top of Dog
Mountain; Pacific Crest Trail at Chini-
dere Mountain.
Saxifrage Family.

214
Parnassia fimbriata   II
var. hoodiana
Fringed Grass-of-   (6-16 in.)
Parnassus
AUG: marshes at middle to high
elevations west of Bonneville Dam.
Closely related *var. fimbriata* blooms
in late August at Wahtum Lake and at
springs along the primitive road (gated)
above the lake.
Saxifrage Family.

215
Saxifraga arguta   II
Brook Saxifrage   (1-2 ft.)
Wet meadows, springs, and stream-
banks at higher elevations in the west
Gorge, on the Oregon side.
MID JULY: North Lake Trailhead;
springs along the primitive road (gated)
north of Wahtum Lake Forest Camp;
Rainy Lake.
Saxifrage Family.

216
Saxifraga bronchialis   III
var. vespertina
Matted Saxifrage   (3-7 in.)
Rocky banks, cliffs, and talus slopes at
either low or high elevations in the
west Gorge.
MID MAY: Scenic Highway between
Oneonta Gorge and Horsetail Falls;
talus slopes near Starvation Creek
State Park.
MID JULY: top of Larch Mountain.
Saxifrage Family.

217
**Saxifraga caespitosa   II
var. subgemmifera**
Tufted Saxifrage   (2-6 in.)
Rocky banks, cliffs, and talus slopes at
either low or high elevations in the
west Gorge.
EARLY MAY: Oneonta Gorge; talus
slopes west of Starvation Creek State
Park.
LATE JUNE-EARLY JULY: top of Larch
Mountain.
Saxifrage Family.

218
**Saxifraga ferruginea   II
var. macounii**
Rusty Saxifrage   (6-14 in.)
Open rocky ground at high elevations
in the west Gorge.
LATE JUNE-EARLY JULY: top of Larch
Mountain; Silver Star Mountain Trail.
LATE JULY: talus slopes on Indian
Mountain.
Saxifrage Family.

219
**Saxifraga integrifolia   V
var. integrifolia**
Northwestern Saxifrage   (6-18 in.)
Moist open slopes throughout the
Gorge, generally at low elevations.
LATE MAR-EARLY APRIL: Memaloose
Viewpoint; SR-14 east of Lyle Tunnel
(MP 77-81).
LATE APRIL-EARLY MAY: SR-14 at the
Clark-Skamania County Line.
Similar *var. claytoniaefolia* also found
throughout the middle and east Gorge.
Saxifrage Family.

220
Saxifraga mertensiana IV
Mertens' Saxifrage (6-16 in.)
Wet shady cliffs as far east as The
Dalles, mostly at low elevations on the
Oregon side.
MID APRIL: Old US-30 east of Hood
River; cliffs along US-30 at Mayer
State Park and at MP 4 on Klickitat
River Road.
EARLY MAY: Scenic Highway between
Multnomah Falls and Oneonta Gorge;
Lower Tanner Creek Road.
Saxifrage Family.

221
Saxifraga occidentalis V
Western Saxifrage (4-10 in.)
Moist banks and cliffs as far east as
The Dalles, generally at low elevations.
LATE MAR: SR-14 at Tunnel #1 and at
Catherine Creek.
MID APRIL: Scenic Highway between
Multnomah Falls and Horsetail Falls.
Saxifrage Family.

222
Saxifraga oregana
var. oregena I
Bog Saxifrage (1-3 ft.)
JULY: Marshes in upper Horsetail
Creek and Upper McCord Creek
Saxifrage Family.

223
Suksdorfia violacea　　　　I
Violet Suksdorfia　　　(4-16 in.)
Moist cliffs at low elevations between
Dog Mountain and The Dalles.
EARLY APRIL: rocks on the west side
of the Klickitat River near the Klickitat
Fish Ladder.
Saxifrage Family.

224
Sullivantia oregana　　　　I
Oregon Sullivantia　　　(3-12 in.)
Wet cliffs near waterfalls at low eleva-
tions in the west Gorge, mostly on the
Oregon side.
EARLY JULY: Multnomah Falls.
Saxifrage Family.

225
Tellima grandiflora　　　　V
Fringe Cup　　　(1-2 ft.)
Moist banks and coniferous woods as
far east as Lyle.
EARLY-MID MAY: Scenic Highway
from Crown Point to Ainsworth Park;
Lower Tanner Creek Road; Dog Creek
Falls.
Saxifrage Family.

226
Tiarella trifoliata IV
var. unifoliata
Coolwort, Foam Flower (8-16 in.)
Woods at middle to high elevations in
the west Gorge.
MID-LATE JULY: Spring Camp Road;
North Lake Trailhead; Wahtum Lake.
Closely related *var. trifoliata* generally
is found toward the west end of the
Gorge.
Saxifrage Family.

227
Tolmiea menziesii V
Youth-on-age (1-2.5 ft.)
Moist woods and streambanks in the
west Gorge.
MID-LATE MAY: Scenic Highway at
Latourell Falls, Bridal Veil Creek, and
Ainsworth State Park; Lower Tanner
Creek Road.
Saxifrage Family.

CURRANT FAMILY

228
Ribes aureum II
Golden Currant (3-9 ft.)
Streambanks and flood plains at low
elevations in the east Gorge.
LATE MAR-EARLY APRIL: Columbia
River bottomlands east of The Dalles
Riverside Park; Bobs Point.
EARLY-MID APRIL: US-30 near Crates
Point; I-84 east of The Dalles.
Currant Family.

229
Ribes bracteosum   III
Stink Currant   (4-9 ft.)
Streambanks and moist woods at all
elevations in the west Gorge.
EARLY MAY: Smith-Cripe Road;
Lower Tanner Creek Road.
LATE MAY: Latourell Falls; Haines
Road at Latourell Creek.
Currant Family.

230
Ribes cereum var. cereum   I
Squaw Currant   (2-6 ft.)
Dry open ground and open woods
from Mayer State Park to the east end
of the Gorge.
EARLY APRIL: SR-14 above Horsethief
Lake.
MID APRIL: US-30 at Mayer State
Park; BLM Parcel at Celilo.
Currant Family.

231
Ribes divaricatum   II
Straggly Gooseberry   (3-6 ft.)
Open woods and thickets, generally at
low elevations in the west Gorge.
LATE MAR: Scenic Highway near
Oneonta Gorge and at Ainsworth
State Park; Sandy River shore.
Currant Family.

232
Ribes lacustre III
Prickly Currant (2-4 ft.)
Moist woods and streambanks, gener-
ally at middle to high elevations in the
west Gorge.
EARLY-MID JUNE: Spring Camp Road;
top of Larch Mountain.
LATE JUNE-EARLY JULY: North Lake
Trailhead; near Indian Springs.
Currant Family.

233
Ribes lobbii I
Pioneer Gooseberry (1.5-3 ft.)
Open woods at middle to high eleva-
tions in the middle Gorge.
EARLY MAY: forest roads near Nestor
Peak.
EARLY JUNE: Monte Carlo Trailhead.
Currant Family.

234
Ribes sanguineum III
Red-flowering Currant (3-9 ft.)
Open woods as far east as The Dalles.
EARLY-MID APRIL: I-84 east of Starva-
tion Creek State Park; Grant Lake.
MID-LATE APRIL: Yeon State Park
Trailhead.
Currant Family.

235
Ribes viscosissimum   II
var. viscosissimum
Sticky Currant   (2-6 ft.)
Dry open woods at higher elevations
in the middle Gorge.
EARLY-MID JUNE: Monte Carlo
Trailhead.
MID-LATE JUNE: Warren Lake
Trailhead.
Currant Family.

HYDRANGEA FAMILY

236
Philadelphus lewisii   IV
Mock Orange   (3-7 ft.)
Cliffs, talus slopes, and watercourses at
low elevations throughout the Gorge.
EARLY JUNE: Old Highway at Rowland
Lake; SR-14 east of Lyle Tunnel (MP
77-81).
MID JUNE: SR-14 at the tunnels
between MP 58-60; US-30 at Mayer
State Park.
Hydrangea Family.

ROSE FAMILY

237
Amelanchier alnifolia   V
Serviceberry   (3-25 ft.)
Open woods and hillsides at all eleva-
tions throughout the Gorge.
EARLY-MID APRIL: Rooster Rock
State Park; Viento State Park; Mema-
loose Rest Area; Dog Creek Falls;
Rowland Lake.
Rose Family.

238
Aruncus sylvester IV
Goat's Beard (2-6 ft.)
Moist woods, especially near streams,
at all elevations in the west Gorge.
EARLY-MID JUNE: Corbett Hill Road;
Scenic Highway at Latourell Falls,
Shepperd's Dell, and Ainsworth State
Park.
Rose Family.

239
Crataegus columbiana I
Columbia Hawthorn (4-8 ft.)
(shown in fruit)
Moist meadows at low elevations east
of The Dalles. Fruits develop by late
July.
LATE APRIL-EARLY MAY: Horsethief
Lake State Park.
Rose Family.

240
Crataegus douglasii II
var. douglasii
Black Hawthorn (6-18 ft.)
Open woods, especially in the middle
Gorge and as far east as Horsethief
Butte, mostly at low elevations.
LATE APRIL-EARLY MAY: Old US-30
west of Mosier; Old Highway near
Major Creek; SR-14 near Gaging
Station; Scenic Highway at Dodson.
Closely-related *var. suksdorfii* is found
in the west Gorge.
Rose Family.

241
Fragaria vesca   V
var. bracteata
Woods Strawberry   (3-7 in.)
Open woods at all elevations in the
west and middle Gorge.
MAY: Scenic Highway at Latourell
Falls, Bridal Veil, and Ainsworth State
Park; Beacon Rock Trail.
Rose Family.

242
Fragaria virginiana   III
var. platypetala
Broad-petal Strawberry   (1-6 in.)
Open woods and meadows as far east
as The Dalles.
LATE APRIL: Eagle Creek Overlook
Picnic Area; Herman Creek Road near
Wyeth.
MID MAY: Carroll Road and Dry
Creek Road south of Mosier.
Rose Family.

243
Geum macrophyllum   III
Large-leaf Avens   (1-2 ft.)
Moist woods and forest openings in
the west and middle Gorge, generally
at low elevations.
EARLY MAY: Scenic Highway at
Latourell Falls, Bridal Veil, and Multno-
mah Falls; Lower Tanner Creek Road.
Rose Family.

244
Holodiscus discolor V
Ocean Spray (3-9 ft.)
Rocky and gravelly soil and cliffs in
open areas, and open woods at all
elevations in the west and middle
Gorge.
MID JUNE: Mayer State Park.
LATE JUNE: cliffs at Viento State Park.
EARLY JULY: I-84 at the Corbett Exit
(#22).
Rose Family.

245
Luetkea pectinata I
Partridge Foot (4-6 in.)
Rocky soil near late snowbanks at high
elevations in the west Gorge.
LATE JULY: Indian Mountain.
Rose Family.

246
Oemleria cerasiformis IV
(Osmaronia cerasiformis)
Indian-plum (4-10 ft.)
Open woods at low elevations in the
west Gorge.
LATE MAR-EARLY APRIL: Ainsworth
State Park; Lower Tanner Creek Road;
SR-14 east of Washougal at MP 20.
Rose Family.

247
Physocarpus capitatus   III
Ninebark   (6-12 ft.)
Moist woods, streambanks, and
marshes at low to middle elevations in
the west Gorge.
LATE MAY: woods along SR-14 east of
Washougal near MP 20.
LATE MAY-EARLY JUNE: Scenic High-
way at Crown Point, Bridal Veil, and
Ainsworth State Park.
Rose Family.

248
Potentilla anserina   III
Common Silverweed   (4-8 in.)
Columbia River shores and bottom-
lands throughout the Gorge.
AUG: Mirror Lake and Young Creek in
Rooster Rock State Park.
LATE AUG-EARLY SEPT: Dalton Point;
mouth of the Deschutes River.
Rose Family.

249
Potentilla drummondii   II
Drummond's Cinquefoil   (6-14 in.)
Meadows at high elevations in the
west Gorge on the Oregon side.
MID JULY: Indian Mountain.
Rose Family.

250
Potentilla flabellifolia I
Fan-leaf Cinquefoil (6-12 in.)
Wet meadows and streambanks at
high elevations.
JULY: Indian Mountain.
Rose Family.

251
Potentilla glandulosa IV
Sticky Cinquefoil (1-2 ft.)
Open woods at all elevations as far
east as The Dalles.
MID-LATE MAY: SR-14 at the Clark-
Skamania County Line and at MP
46.6; Buck Creek Road.
EARLY-MID JUNE: Grassy Knoll
Trailhead.
Rose Family.

252
Potentilla gracilis II
Graceful Cinquefoil (1-2 ft.)
Moist open areas and open woods at
low to middle elevations as far east as
Horsethief Butte.
EARLY JUNE: base of Cape Horn (near
the river); Gorge Trail about one mile
east of Multnomah Falls.
LATE JUNE: meadows on the Major
Creek Plateau; Horsethief Butte.
Rose Family.

253
Potentilla rivalis II
Riverbank Cinquefoil (4-14 in.)
Columbia River shores and other low
elevation wet areas throughout the
Gorge.
JULY: west bank of the Deschutes
River between I-84 and the railroad
tracks.
MID SEPT: Dalton Point; Columbia
River shore west of Rooster Rock.
Rose Family.

254*
Prunus cerasus III
Sour Cherry (15-30 ft.)
Low elevation woods in the west
Gorge.
MID APRIL: Scenic Highway at
Ainsworth State Park; Starvation
Creek State Park.
MID MAY: Larch Mountain Highway.
Rose Family.

255
Prunus emarginata III
var. emarginata
Bitter Cherry (3-12 ft.)
Open areas and open woods at all
elevations between Cascade Locks and
Horsethief Butte.
LATE APRIL: SR-14 at Wind Mountain,
Dog Mountain Trailhead, and near MP
70.4; I-84 at Mayer State Park.
Closely-related *var. mollis* is a small
tree (6-40 ft.) found at low elevations
in the west Gorge.
Rose Family.

256
Prunus virginiana   II
var. melanocarpa
Choke Cherry   (6-12 ft.)
Open slopes and open woods at low
elevations from Beacon Rock to the
east end of the Gorge.
EARLY-MID MAY: Rowland Lake;
Horsethief Butte; The Dalles Mountain
Road.
Similar *var. demissa* is found in the
west Gorge, blooming in the last half
of May.
Rose Family.

257
Purshia tridentata   III
Bitter Brush   (2-6 ft.)
Dry open areas east of Hood River,
mostly at low elevations.
LATE APRIL: Horsethief Butte; Bobs
Point; The Dalles Bridge Road (US-
197); US-30 at Crates Point.
LATE MAY: near the top of Chenoweth
Road.
Rose Family.

258
Pyrus fusca   II
Western Crab Apple   (9-30 ft.)
Streambanks, lake shores, and marshes
in the west Gorge.
LATE APRIL: Wind River shores above
the mouth of the Little Wind River;
Berge Road just past MP 1.
EARLY MAY: Gorge Trail at Moffett
Creek.
Rose Family.

259*
Rosa canina II
Dog Rose (3-6 ft.)
Roadsides and disturbed areas as far
east as The Dalles.
LATE MAY-EARLY JUNE: I-84 near
Rowena at MP 77.
MID JUNE: Corbett Hill Road.
Rose Family.

260
Rosa gymnocarpa IV
Little Wild Rose (1.5-4 ft.)
Open woods at all elevations in the
west and middle Gorge.
LATE MAY: Latourell Falls; Mt. Zion;
Beacon Rock; Grant Lake.
Rose Family.

261
Rosa nutkana IV
Nootka Rose (3-6 ft.)
Open woods at all elevations as far
east as The Dalles.
LATE MAY: Scenic Highway at Bridal
Veil, Multnomah Falls, and Ainsworth
State Park; SR-14 west of Beacon
Rock; Rock Creek Lake; Grant Lake.
Rose Family.

262
Rosa pisocarpa   II
Clustered Wild Rose   (3-6 ft.)
Open woods and damp open ground
at low to middle elevations in the west
and middle Gorge.
MID-LATE JUNE: Multnomah Falls;
Home Valley Park; Grant Lake.
Rose Family.

263
Rosa woodsii   II
var. ultramontana
Pearhip Rose   (3-6 ft.)
Moist open ground and open woods in
the east Gorge.
LATE MAY: SR-14 at Tunnel #1, east
of Lyle Tunnel (MP 77-81), and at
Horsethief Butte; US-30 at Chenoweth
Creek (MP 15).
Rose Family.

264 *
Rubus discolor   V
(Rubus procerus)
Himalayan Blackberry   (3-8 ft.)
Roadsides, railroads, pastures, and
other disturbed open land.
LATE JUNE: Rowland Lake.
LATE JUNE-EARLY JULY: Scenic High-
way at Bridal Veil Falls State Park (MP
15.6).
MID JULY: Dalton Point; Larch Moun-
tain Highway.
Rose Family.

265
Rubus lasiococcus   IV
Dwarf Bramble   (1-4 in.)
Higher elevation woods in the west
Gorge.
EARLY JULY: Spring Camp Road; top
of Larch Mountain; Rainy Lake.
Rose Family.

266
Rubus leucodermis   III
Blackcap   (3-7 ft.)
Open ground and open woods at low
to middle elevations as far east as The
Dalles.
EARLY MAY: SR-14 at Wind Mountain.
MID MAY: I-84 east of Multnomah
Falls.
EARLY JUNE: Brower Road.
Rose Family.

267
Rubus parviflorus   V
Thimbleberry   (2-6 ft.)
Open woods and slopes at all eleva-
tions in the west and middle Gorge.
MID MAY: I-84 in the west Gorge;
Scenic Highway; SR-14 west of Beacon
Rock.
Rose Family.

268
Rubus pedatus   II
Five-leaf Bramble   (2-4 in.)
Moist woods at higher elevations as far east as Mt. Defiance.
LATE MAY-EARLY JUNE: woods along Larch Mountain Highway near the Mt. Hood National Forest boundary.
LATE JUNE-EARLY JULY: top of Larch Mountain; trails in Multnomah Basin at about 3,000 ft. elev.
Rose Family.

269
Rubus spectabilis   IV
Salmonberry   (3-7 ft.)
Open moist woods and creek bottoms at all elevations in the west Gorge.
EARLY APRIL: Latourell Falls; Lower Tanner Creek Road.
EARLY JUNE: top of Larch Mountain.
Rose Family.

270
Rubus ursinus   IV
Trailing Blackberry   (vine)
Clearings and open woods at all elevations in the west and middle Gorge.
EARLY MAY: SR-14 at Dog Mountain.
MID MAY: Scenic Highway at Latourell Falls and Ainsworth State Park.
Rose Family.

271
Sanguisorba officinalis   II
Marsh Burnet   (1-2.5 ft.)
AUG: wet meadows and marshes at
middle to high elevations west of
Bonneville Dam.
Rose Family.

272
Sorbus scopulina   III
Cascade Mountain Ash   (3-12 ft.)
Open woods at higher elevations in
the west Gorge.
MID-LATE JUNE: top of Larch Moun-
tain.
Rose Family.

273
Sorbus sitchensis   II
Sitka Mountain Ash   (3-12 ft.)
Higher elevations in the west Gorge.
EARLY-MID JULY: Indian Mountain.
Rose Family.

274
Spiraea betulifolia   IV
var. lucida
White Spiraea   (1-2 ft.)
Dry open woods at all elevations as far
east as The Dalles.
MID JUNE: Bridal Veil Falls State Park.
LATE JUNE: SR-14 west of Cape Horn.
EARLY AUG: top of Larch Mountain.
Rose Family.

275
Spiraea densiflora   II
Mountain Spiraea   (1-2 ft.)
Open forests and meadows at high
elevations in the west Gorge.
LATE JUNE-EARLY JULY: near the top
of Silver Star Mountain.
MID-LATE JULY: near Indian Springs.
Rose Family.

276
Spiraea douglasii   III
Western Spiraea,   (2-6 ft.)
Steeple Bush
Ditches, wet meadows, lake margins,
and Columbia River bottomlands at all
elevations as far east as Horsethief
Butte.
LATE JUNE-EARLY JULY: SR-14 at
Horsethief Butte.
LATE JULY-EARLY AUG: Young Creek
in Rooster Rock State Park; Home
Valley Park; Beacon Rock Pond.
Rose Family.

277
Spiraea pyramidata   I
Pyramid Spiraea   (2-4 ft.)
Open woods at scattered locations in
the west Gorge.
EARLY-MID JUNE: SR-14 near MP 20;
Hamilton Mountain Trail above Hardy
Falls.
Rose Family.

PEA FAMILY

278
Amorpha fruticosa   III
Western False Indigo   (6-10 ft.)
Columbia River shores throughout the
Gorge.
MID JUNE: Columbia River shore just
west of Rooster Rock; Dalton Point;
Beacon Rock shore; Home Valley Park.
Pea Family.

279
Astragalus hoodianus   III
Hood River Milk-vetch   (1-2 ft.)
Open or lightly wooded areas at low to
middle elevations between the White
Salmon River and Horsethief Butte.
LATE APRIL-EARLY MAY: US-30 at
the west edge of Mayer State Park
(MP 6.5); SR-14 near the junction with
The Dalles Bridge Road (US-197).
MID MAY: Hood River Mountain
Meadow.
Pea Family.

280
Astragalus purshii III
Woolly-pod Milk-vetch (1-3 in.)
Open grassy slopes in the east Gorge.
LATE MAR-EARLY APRIL: Wishram
Historical Marker.
MID APRIL: The Dalles Mountain
Road; Dry Creek Road at MP 3.5.
Pea Family.

281
Astragalus reventiformis II
Yakima Milk-vetch (4-14 in.)
Open slopes on the north side of the
Columbia River from Horsethief Butte
to the east end of the Gorge.
MID-LATE APRIL: The Dalles Mountain
Road.
EARLY MAY: Haystack Butte Road.
Pea Family.

282
Astragalus sclerocarpus I
The Dalles Milk-vetch (10-20 in.)
Sparse populations found at low
elevations east of The Dalles.
EARLY MAY: I-84 (eastbound) on the
hillside near the turnout at MP 92.3.
MID APRIL: east of Avery Gravel Pit.
Pea Family.

283
Astragalus succumbens   I
Columbia Milk-vetch   (4-8 in.)
Open slopes at low elevations near
Wishram; thus far found only on the
Washington side.
LATE APRIL-EARLY MAY: slopes
above SR-14 east of Wishram, espe-
cially at MP 91, 94, and 95, and in the
highway right-of-way at MP 95.9.
Pea Family.

284
Glycyrrhiza lepidota   III
var. glutinosa
Licorice Root   (1-4 ft.)
Columbia River shores throughout the
Gorge; occasionally in wet places up
from the river.
MID JULY-MID AUG: Dalton Point;
Bobs Point.
Pea Family.

285
Hedysarum occidentale   I
Western Sweet-vetch   (1.5-3 ft.)
EARLY-MID JULY: thus far found in
the Gorge only near the top of Silver
Star Mountain and adjacent rocky
peaks.
Pea Family.

286*
Lathyrus latifolius III
Everlasting Pea (2-4 ft.)
Roadsides and railroad embankments
in the west Gorge.
EARLY JULY: I-84 at Viento State Park;
SR-14 from the White Salmon River to
Bingen.
MID JULY: Scenic Highway at Crown
Point; SR-14 west of Stevenson.
Pea Family.

287
Lathyrus nevadensis IV
var. pilosellus
Nevada Pea (1-2.5 ft.)
Open woods in the west Gorge.
LATE APRIL: SR-14 at the base of Dog
Mountain.
LATE MAY: Scenic Highway at Multno-
mah Falls.
Pea Family.

288
Lathyrus pauciflorus III
var. pauciflorus
Few-flowered Pea (1-2 ft.)
Open woods and grasslands in the east
Gorge.
LATE APRIL: Skyline Road south of
The Dalles.
EARLY-MID MAY: Old US-30 east of
Hood River.
MID MAY: roads near Wasco Butte.
Pea Family.

289
Lathyrus polyphyllus IV
Leafy Pea (1.5-3 ft.)
Open woods in the west and middle
Gorge.
LATE MAY-EARLY JUNE: Scenic High-
way at Bridal Veil and at Ainsworth
State Park.
Pea Family.

290*
Lotus corniculatus II
Bird's-foot Trefoil (1-2 ft.)
Damp areas at low to middle elevations
throughout the Gorge, particularly on
the Washington side.
LATE JUNE-EARLY JULY: SR-14 near
Bonneville Dam.
MID JULY: Wind River shore; ditches
on the Major Creek Plateau.
Pea Family.

291
Lotus crassifolius I
var. crassifolius
Thick-leaf Deer-vetch (2-4 ft.)
Open woods in the middle Gorge.
MID JUNE: upper edge of Hood River
Mountain Meadow.
Pea Family.

292
Lotus crassifolius II
var. subglaber
Thicket Deer-vetch (1-2 ft.)
Open woods and clearings in the west
end of the Gorge.
MID JUNE: Larch Mountain Corridor;
Brower Road.
Pea Family.

293
Lotus micranthus III
Small-flowered Deer-vetch (4-10 in.)
Open woods and grasslands, mostly in
the middle Gorge.
MAY: SR-14 at Wind Mountain;
grassy slopes above Rowland Lake.
Pea Family.

294
Lotus nevadensis III
Nevada Deer-vetch (2-8 in.)
Dry, often rocky forest openings,
mostly between Dog Mountain and
The Dalles.
LATE MAY-EARLY JUNE: Old Highway
at Rowland Lake and at the gravel pit
just east of Major Creek.
MID JUNE: SR-14 at the base of Dog
Mountain.
EARLY JULY: Hood River Mountain
Meadow; top of Angels Rest.
Pea Family.

295
Lotus pinnatus II
Bog Deer-vetch (8-18 in.)
Slow-moving water, often where dry
by late summer, at low to middle
elevations in the west and middle
Gorge.
MID MAY: Campbell Creek.
MID-LATE JUNE: ditch along Bristol
Road at Laws Corner.
Pea Family.

296
Lotus purshiana III
Spanish-clover (8-18 in.)
Dry places at low elevations through-
out the Gorge.
SUMMER-FALL: I-84 at Rooster Rock
State Park; US-30 at Mayer State Park;
SR-14 east of Bingen.
Pea Family.

297
Lupinus bicolor III
Bicolored Lupine (4-10 in.)
Dry places at low elevations between
the Wind River and The Dalles.
MID APRIL: US-30 near Rowena; SR-
14 west of Lyle Tunnel.
MAY: SR-14 west of Wind Mountain
and east of Dog Mountain.
Pea Family.

298
Lupinus latifolius   V
var. latifolius
Common Broad-leaf   (1.5-3.5 ft.)
Lupine
Open areas and open woods at all
elevations in the west Gorge.
MAY: I-84 from Starvation Creek State
Park to Hood River; SR-14 from Dog
Mountain to the White Salmon River.
JUNE-JULY: I-84 at Cascade Locks and
Wyeth; Larch Mountain Highway.
Pea Family.

299
Lupinus latifolius   IV
var. thompsonianus
Columbia Gorge   (1.5-3 ft.)
Broad-leaf Lupine
Open woods and grasslands in the east
Gorge, mainly on the Washington side.
LATE APRIL: Old Highway at Rowland
Lake; SR-14 at the junction with The
Dalles Bridge Road (US-197).
LATE MAY: The Dalles Mountain
Road, especially at the top.
Pea Family.

300
Lupinus laxiflorus   IV
var. laxiflorus
Spurred Lupine   (1.5-2.5 ft.)
Open woods and grasslands at all
elevations from Bonneville Dam to the
east end of the Gorge.
MID MAY: near Catherine Creek.
LATE MAY-EARLY JUNE: I-84 near
Starvation Creek State Park, near
Viento State Park, and west of Mosier.
MID JUNE: Hood River Mountain
Meadow.
Pea Family.

301
Lupinus lepidus III
var. aridus
Prairie Lupine (4-8 in.)
Dry open ground between Lyle and
Wishram, mostly at low elevations on
the Washington side.
LATE MAY-EARLY JUNE: SR-14 be-
tween Lyle Tunnel and the junction
with The Dalles Bridge Road (US-197);
Horsethief Lake State Park.
Pea Family.

302
Lupinus leucophyllus II
Velvet Lupine (1-2.5 ft.)
Dry open ground at all elevations east
of Lyle.
JUNE: I-84 between Rowena and
Crates Point; SR-14 near the junction
with The Dalles Bridge Road (US-197);
The Dalles Mountain Road.
Pea Family.

303
Lupinus leucopsis IV
(Lupinus sulphureus
var. subsaccatus)
Whitish Lupine (1-2.5 ft.)
Open ground at all elevations in the
east Gorge.
MID-LATE APRIL: SR-14 near the rest
area at MP 74 and east of Lyle Tunnel
(MP 77-81).
MAY: large stands of this lupine ''blue''
the hillsides north of Dallesport and
along I-84 east of The Dalles.
Pea Family.

304
Lupinus micranthus   III
Small-flowered Lupine   (4-10 in.)
Open areas throughout the Gorge,
generally at low elevations.
LATE APRIL-EARLY MAY: I-84 at
Cascade Locks; Memaloose Viewpoint;
Catherine Creek.
Pea Family.

305
Lupinus polyphyllus   III
Large-leaf Lupine   (2-4 ft.)
Moist ground in the west and middle
Gorge.
MID MAY-MID JUNE: SR-14 between
Washougal and Beacon Rock.
JUNE: I-84 at Moffett Creek; bluff at
Bridal Veil Falls State Park; meadows
on the Major Creek Plateau.
Pea Family.

306
Lupinus rivularis   I
Riverbank Lupine   (1.5-3.5 ft.)
LATE APRIL-EARLY MAY: a shrubby
species, thus far found in the Gorge
only along I-84 at the east end of
Rooster Rock State Park, near MP
25-26.
Pea Family.

307
Lupinus saxosus   I
Stony-ground Lupine   (6-10 in.)
Dry open ground at higher elevations
in The Dalles area. Thus far found only
on the Oregon side.
LATE MAR-EARLY APRIL: near Seven-
Mile Hill Road at about 1,400 ft. elev.
MID APRIL: near the top of Cheno-
weth Road.
Pea Family.

308
Lupinus sericeus   II
Silky Lupine   (8-24 in.)
Dry open slopes in the east Gorge,
where it "blues" the hillsides around
the first of June.
JUNE-JULY: SR-14 from Wishram to
Maryhill.
Pea Family.

309*
Melilotus alba   III
White Sweet-clover   (2-7 ft.)
Roadsides and waste places through-
out the Gorge, mostly at low eleva-
tions.
EARLY-MID JULY: SR-14 from Dog
Mountain to the White Salmon River;
Lower Tanner Creek Road; Crown
Point.
Pea Family.

310
Psoralea lanceolata   III
Lance-leaf Scurf-pea   (6-16 in.)
Sandy ground east of Mosier, particu-
larly on the windward side of active
dunes.
JUNE-JULY: dunes near The Dalles
Airport; Horsethief Butte; sandy places
along I-84 east of The Dalles.
Pea Family.

311
Thermopsis montana   I
var. ovata
Golden-pea   (1.5-3 ft.)
Open places and roadsides at middle
to high elevations in the west end of
the Gorge.
JUNE-JULY: near Silver Star Mountain;
North Fork Washougal River Road.
Pea Family.

312*
Trifolium arvense   III
Rabbit-foot Clover   (6-16 in.)
Roadsides and disturbed areas at low
elevations as far east as The Dalles.
LATE JUNE-EARLY JULY: I-84 east of
the Sandy River, at Rooster Rock State
Park, and near Starvation Creek State
Park.
Pea Family.

313
Trifolium ciliolatum II
Tree Clover (6-18 in.)
Open grasslands at low elevations
between Bingen and The Dalles.
MID MAY: Locke Lake; Tom McCall
Nature Preserve.
Pea Family.

314
Trifolium cyathiferum II
Cup Clover (4-14 in.)
Open areas at low to middle elevations
between Cascade Locks and The
Dalles.
LATE MAY-EARLY JUNE: Tom McCall
Nature Preserve; SR-14 near the rest
area west of Lyle (MP 74).
Pea Family.

315
Trifolium eriocephalum I
var. eriocephalum
Woolly-head Clover (8-18 in.)
EARLY MAY: open woods; thus far
found in the Gorge only along Mosier
Creek Road, about two miles south of
Mosier.
Pea Family.

316
Trifolium macraei I
var. dichotomum
Purple Clover (4-14 in.)
EARLY MAY: thus far found in the
Gorge only on a grassy bench above
SR-14 just east of Bingen.
Pea Family.

317
Trifolium macrocephalum III
Big-head Clover (3-8 in.)
Grassy areas and open woods at all
elevations, mainly between Dog
Mountain and The Dalles.
APRIL: backroads between Mosier and
The Dalles; Old Highway; Major Creek
Road.
MAY: roads near Wasco Butte.
Pea Family.

318
Trifolium microcephalum II
Small-head Clover (3-12 in.)
Open areas at low to middle elevations,
primarily between Dog Mountain and
The Dalles.
MAY: SR-14 near the junction with
The Dalles Bridge Road (US-197) and
near the rest area west of Lyle (MP
74); US-30 between Mosier and
Mayer State Park.
Pea Family.

319
Trifolium tridentatum III
Sand Clover (6-14 in.)
Open grasslands at low elevations,
mostly between Wind Mountain and
Lyle.
LATE APRIL-EARLY MAY: Locke Lake;
slope above SR-14 at MP 70.4; Old
Highway near Catherine Creek.
Pea Family.

320
Trifolium variegatum II
White-tip Clover (4-14 in.)
Open ground at all elevations, mainly
between the White Salmon River and
The Dalles.
MAY: Memaloose Viewpoint; Tom
McCall Nature Preserve; Old Highway
near Catherine Creek.
Pea Family.

321
Vicia americana V
var. truncata
American Vetch (6-30 in.)
Open woods and roadsides at all
elevations in the west and middle
Gorge.
MAY: US-30 west and east of Mosier;
Old Highway near Rowland Lake.
LATE MAY-EARLY JUNE: Larch Moun-
tain Highway near MP 4.
Pea Family.

322
Vicia gigantea I
Giant Vetch (1-3 ft.)
Low elevations in the west end of the
Gorge; thus far found only on the
Oregon side.
EARLY JUNE: Scenic Highway near
Bridal Veil and near Crown Point.
Pea Family.

323*
Vicia sativa IV
Common Vetch (1-2 ft.)
Roadsides, fields, and disturbed areas
as far east as The Dalles.
LATE MAY: SR-14 at Wind Mountain,
Grant Lake, and Dog Mountain; I-84
at Rooster Rock State Park and Viento
State Park.
Pea Family.

324*
Vicia villosa IV
Annual Cow Vetch (1-3 ft.)
Roadsides and disturbed areas at low
elevations throughout the Gorge.
MID JUNE: I-84 at Rooster Rock State
Park, Viento State Park, and Rowena;
SR-14 at Wind Mountain, Grant Lake,
and from the White Salmon River to
Bingen.
Pea Family.

SECTION THREE

GERANIUM FAMILY

325*
Erodium cicutarium V
Filaree (1-12 in.)
Dry open ground throughout the
Gorge, but especially in the east Gorge.
EARLY APRIL: SR-14 east of Lyle
Tunnel (MP 77-81).
LATE APRIL: I-84 at Multnomah Falls.
Geranium Family.

326
Geranium carolinianum II
Carolina Geranium (3-16 in.)
Moist open areas, mainly at low eleva-
tions between Hood River and The
Dalles, often near oaks.
LATE MAY: Spring Creek Fish
Hatchery.
Geranium Family.

WOOD SORREL FAMILY

327
Oxalis oregana III
Oregon Wood Sorrel (4-6 in.)
Moist woods at low to middle eleva-
tions as far east as Cascade Locks.
APRIL: Scenic Highway at Latourell
Falls and Ainsworth State Park.
Wood Sorrel Family.

328
Oxalis trilliifolia　　　　　II
Trillium-leaf Wood Sorrel　　(5-10 in.)
Moist woods at all elevations as far
east as Cascade Locks.
JUNE-JULY: Scenic Highway near
Crown Point and at Latourell Falls.
Wood Sorrel Family.

SUMAC FAMILY

329
Rhus diversiloba　　　　　V
Poison Oak　　　　　　　(1-8 ft.)
Forest openings and talus slopes at low
elevations as far east as The Dalles;
often an extensive ground cover in the
shade of conifers and oaks.
LATE MAY-EARLY JUNE: Viento State
Park; Stanley Rock.
Sumac Family.

330
Rhus glabra　　　　　II
Western Sumac　　　　(3-10 ft.)
Shrub or small tree found at low
elevations east of The Dalles.
EARLY JULY: SR-14 near MP 89.5;
east bank of the Deschutes River.
Sumac Family.

331
Rhus radicans   III
Poison Oak (shown in fruit)   (1-6 ft.)
Dry ground and talus slopes at low
elevations, mainly east of The Dalles.
LATE MAY-EARLY JUNE: Horsethief
Butte; DNR Falls.
Sumac Family.

STAFF-TREE FAMILY

332
Pachistima myrsinites   II
Oregon Boxwood   (1-2 ft.)
Forest openings, mostly at high eleva-
tions in the west Gorge.
APRIL: Beacon Rock Trail.
MID JUNE: top of Larch Mountain.
Staff-tree Family.

BALSAM FAMILY

333
Impatiens capensis   I
Orange Balsam   (1-3.5 ft.)
Columbia River shores and bottom-
lands in the west Gorge.
MID AUG: south shore of Bradford
Island.
Balsam Family.

334
Ceanothus integerrimus   IV
Deer Brush   (3-12 ft.)
Open areas and forest openings
between Dog Mountain and The
Dalles. Colors range from white to
deep blue.
LATE MAY-EARLY JUNE: SR-14 be-
tween Dog Mountain and the Hood
River Bridge; Mayer State Park.
Buckthorn Family.

335
Ceanothus sanguineus   III
Buck Brush   (3-9 ft.)
Open woods and thickets at low to
middle elevations in the west and
middle Gorge.
EARLY-MID MAY: I-84 at Wyeth,
Starvation Creek State Park, Viento
State Park, Mitchell Point, and Hood
River.
MID MAY: Scenic Highway.
Buckthorn Family.

336
Ceanothus velutinus   II
Snow Brush   (3-6 ft.)
Dry open woods and clearings at
middle to high elevations in the west
Gorge.
LATE JUNE: Warren Lake Road; Monte
Carlo Trailhead.
Buckthorn Family.

337
Iliamna rivularis   II
Mountain Hollyhock   (2-6 ft.)
Open forests as far east as the Major
Creek Plateau; thus far found only on
the Washington side.
MID JUNE-MID JULY: Buck Creek
Road; Snowden Road near the Bates
Road junction.
LATE JULY-EARLY AUG: top of Monte
Carlo.
Mallow Family.

338
Sidalcea oregana   II
Marsh Hollyhock   (1.5-4 ft.)
Seasonally wet areas such as stream
margins, meadows, and vernal ponds
between Hood River and Horsethief
Lake.
LATE MAY-EARLY JUNE: Campbell
Creek.
LATE JUNE-EARLY JULY: meadows on
the Major Creek Plateau.
Mallow Family.

ST. JOHN'S WORT FAMILY

339
Hypericum anagalloides   II
Bog St John's Wort   (1-3 in.)
Marshes and lake margins at middle to
high elevations in the west and middle
Gorge.
LATE JULY-EARLY AUG: east shore of
Wahtum Lake.
St. John's Wort Family.

340
Hypericum formosum   I
Western St. John's Wort   (1-2.5 ft.)
Marshes, wet meadows, and ditches at
middle to high elevations in the middle
Gorge.
MID-LATE JULY: ditches along Bristol
Road south of Laws Corner; meadows
on the Major Creek Plateau.
St. John's Wort Family.

341*
Hypericum perforatum   V
Common St. John's Wort   (1-2 ft.)
Fields, roadsides, and adjacent areas at
all elevations throughout the Gorge.
LATE JUNE: SR-14 between Dog
Mountain and Bingen.
MID JULY: Scenic Highway near
Crown Point.
St. John's Wort Family.

VIOLET FAMILY

342
Viola adunca   II
Long-spurred Violet   (3-6 in.)
Meadows and open forests, generally
at middle to high elevations in the west
and middle Gorge.
EARLY MAY: near the top of Hamilton
Mountain Trail.
JULY: Chinidere Mountain.
Violet Family.

343
Viola glabella V
Stream Violet (4-6 in.)
Moist woods and streambanks at all
elevations in the west and middle
Gorge.
LATE APRIL: Scenic Highway at La-
tourell Falls, Bridal Veil, and Ainsworth
State Park; Lower Tanner Creek Road.
Violet Family.

344
Viola howellii II
Howell's Violet (4-6 in.)
Woods and clearings at middle eleva-
tions in the west and middle Gorge,
mostly on the Washington side.
LATE APRIL: Berge Road just past
MP 1.
MAY: Laws Corner.
Violet Family.

345
Viola nuttallii III
Yellow Prairie Violet (4-12 in.)
Open woods, generally at middle to
high elevations east of Dog Mountain.
LATE MAR: along SR-14 at the west
edge of Bingen.
LATE APRIL: Campbell Creek; Laws
Corner.
Violet Family.

346
Viola orbiculata   I
Round-leaf Violet   (4-8 in.)
Higher elevations in the west Gorge.
LATE MAY-EARLY JUNE: top of Larch
Mountain.
Violet Family.

347
Viola palustris   II
Marsh Violet   (2-6 in.)
Marshes at higher elevations in the
west Gorge; occasionally on stream-
banks at lower elevations.
LATE APRIL-EARLY MAY: Wind River
shore.
LATE JUNE-EARLY JULY: spring near
North Lake Trailhead; Rainy Lake.
Violet Family.

348
Viola sempervirens   III
Evergreen Violet   (2-5 in.)
Coniferous woods at all elevations in
the west Gorge.
MID APRIL: Beacon Rock Trail.
MID JUNE: top of Larch Mountain.
Violet Family.

349
Viola sheltonii   I
Cut-leaf Violet   (2-6 in.)
Scattered populations on the Washing-
ton side between the Wind River and
the Klickitat River, usually in the partial
shade of oaks.
EARLY-MID APRIL: grassy, open or
lightly wooded slopes below Major
Creek Road.
Violet Family.

BLAZING STAR FAMILY

350
Mentzelia laevicaulis   II
var. laevicaulis
Blazing Star   (1.5-3 ft.)
Open, gravel or talus slopes at low
elevations, mainly on the Washington
side between Dog Mountain and The
Dalles area.
LATE JULY: SR-14 at the base of Dog
Mountain, at the base of cliffs east of
Bingen near MP 69, and east of Lyle
Tunnel (MP 77-81).
Blazing Star Family.

CACTUS FAMILY

351
Opuntia polyacantha   II
Prickly-pear Cactus   (4-8 in.)
Flat or gently sloping open ground at
low elevations east of Lyle Tunnel on
the Washington side.
EARLY JUNE: SR-14 near the turnoff
to the Avery Gravel Pit (MP 89.7);
SR-14 at MP 80.5 and at MP 91.
Cactus Family.

352
Ammannia coccinea   I
Scarlet Ammannia   (4-12 in.)
Muddy shores of the Columbia River.
EARLY SEPT: Columbia River shore
east of The Dalles Riverside Park.
Loosestrife Family.

353 *
Lythrum salicaria   I
Purple Loosestrife   (3-6 ft.)
Roadside ditches and the margins of
lakes and streams between the Wind
River and Dog Mountain.
LATE JULY: SR-14 at Home Valley and
Grant Lake; Home Valley Park.
Loosestrife Family.

354
Boisduvalia densiflora   III
Dense Spike-primrose   (6-36 in.)
Margins of vernal ponds and streams
on both sides of the Gorge between
Grant Lake and Horsethief Butte.
MID JULY: Tom McCall Nature Pre-
serve; Horsethief Butte; wet meadows
and ditches on the Major Creek
Plateau.
Evening-primrose Family.

355
Boisduvalia glabella II
Smooth Spike-primrose (6-20 in.)
Margins of vernal ponds in the east
Gorge.
LATE JUNE-EARLY JULY: Tom McCall
Nature Preserve; Horsethief Lake State
Park.
Evening-primrose Family.

356
Boisduvalia stricta II
Stiff Spike-primrose (6-18 in.)
Margins of vernal ponds and streams
between Bingen and Horsethief Lake
State Park, blooming when the soil
becomes nearly dry.
MID JULY: ditches and draws along
the Old Highway; Tom McCall Nature
Preserve.
Evening-primrose Family.

357
Circaea alpina III
Enchanter's Nightshade (4-14 in.)
Moist coniferous woods in the west
and middle Gorge.
JUNE: Scenic Highway at Latourell
Falls, Shepperd's Dell, and Bridal Veil
Creek.
Evening-primrose Family.

358
Clarkia amoena III
var. lindleyi
Herald-of-summer (6-24 in.)
Open, grassy or rocky areas in the
west Gorge, often near oaks.
MID JUNE: SR-14 at Signal Rock and
Tunnel #1.
LATE JUNE-EARLY JULY: SR-14 at the
Clark-Skamania County Line.
The closely related *var. caurina* is also
found in the west Gorge.
Evening-primrose Family.

359
Clarkia gracilis II
Slender Godetia (6-18 in.)
Open grassy areas at all elevations
between the White Salmon River and
The Dalles.
MID MAY: Major Creek Road; Mema-
loose Viewpoint.
MID JUNE: Hood River Mountain
Meadow.
Evening-primrose Family.

360
Clarkia pulchella I
Elkhorns Clarkia (10-20 in.)
Open, grassy or brushy areas between
Hood River and The Dalles.
LATE JUNE-EARLY JULY: Tom McCall
Nature Preserve.
Evening-primrose Family.

361
Clarkia quadrivulnera III
Small-flowered Godetia (6-24 in.)
Open grassy places and lightly wooded
areas at all elevations between Dog
Mountain and Horsethief Lake State
Park.
MID JUNE: Tom McCall Nature Pre-
serve; Old Highway near Catherine
Creek; Major Creek Road.
Evening-primrose Family.

362
Clarkia rhomboidea II
Common Clarkia (6-20 in.)
Dry forest openings at low to middle
elevations between Bonneville Dam
and the Klickitat River.
MID JUNE: lower part of Dog Moun-
tain Trail; upper Rock Creek Road.
Evening-primrose Family.

363
Epilobium alpinum II
var. lactiflorum
White-flowered (6-10 in.)
Willow Herb
Open woods at middle to high eleva-
tions in the west Gorge.
LATE JUNE-EARLY JULY: Spring Camp
Road.
EARLY JULY: top of Larch Mountain.
Evening-primrose Family.

364
Epilobium alpinum   I
var. nutans
Alpine Brook Willow Herb   (6-10 in.)
Springs and streambanks in subalpine
forests at over 3,500 ft. elev. in the
west Gorge.
MID JULY: springs along the primitive
road (gated) north of Wahtum Lake
Forest Camp.
Evening-primrose Family.

365
Epilobium angustifolium   IV
Fireweed   (3-10 ft.)
Roadsides, burns, disturbed areas, and
other forest openings at all elevations,
mainly in the west Gorge.
JULY: scattered locations along I-84,
Scenic Highway, and Larch Mountain
Highway.
AUG: top of Larch Mountain.
Evening-primrose Family.

366
Epilobium ciliatum   V
subspecies ciliatum
(Epilobium watsonii
var. occidentale)
Common Willow Herb   (1-3 ft.)
Wet places throughout the Gorge,
generally at low elevations.
MID-LATE JULY: ditches on SR-14 east
of Cape Horn.
LATE AUG-EARLY SEPT: Columbia
River shores at Rooster Rock State
Park, Dalton Point; Mirror Lake.
Evening-primrose Family.

367
Epilobium ciliatum   II
subspecies glandulosum
(Epilobium glandulosum)
Bog Willow Herb   (1-3 ft.)
Wet meadows and marshy shores of
lakes and streams at higher elevations
in the west Gorge.
LATE JULY: stream at North Lake
Trailhead.
EARLY AUG: springs along the primi-
tive road (gated) above Wahtum Lake.
Evening-primrose Family.

368*
Epilobium hirsutum   II
Fiddle-grass   (2.5-4 ft.)
Ditches, riverbanks, and other wet
areas at low elevations from the Wind
River to the east end of the Gorge.
AUG: Home Valley Park; SR-14 west
and east of Bingen and at Tunnel #1;
west bank of the Deschutes River.
Evening-primrose Family.

369
Epilobium luteum   I
Yellow Willow Herb   (8-14 in.)
Streambanks at higher elevations in
the west Gorge.
AUG: thus far found in the Gorge only
at Three-Creek Camp on Eagle Creek
Trail at 3,300 ft. elev.
Evening-primrose Family.

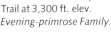

370
Epilobium minutum   III
Small-flowered   (4-12 in.)
Willow Herb
Dry, often rocky areas at all elevations
as far east as The Dalles.
LATE APRIL: Dog Creek Falls; Old
Highway at Catherine Creek.
MID-LATE MAY: Beacon Rock Trail.
Evening-primrose Family.

371
Epilobium paniculatum   III
var. jucundum
Large-flowered Annual   (1-3 ft.)
Willow Herb
Open grasslands, roadsides, and other
dry places in the east Gorge.
AUG-SEPT: backroads between Mosier
and The Dalles; Bobs Point; Horsethief
Lake State Park.
Evening-primrose Family.

372
Epilobium paniculatum   IV
var. paniculatum
Tall Annual Willow Herb   (1-3 ft.)
Dry open places, mostly at low eleva-
tions; often on road embankments and
other disturbed areas throughout the
Gorge.
AUG: Lower Tanner Creek Road;
Grant Lake.
Evening-primrose Family.

373
Gaura parviflora　　　　　　I
Small-flowered Gaura　　　(2-6 ft.)
A few plants at scattered locations east
of The Dalles.
JUNE: near the old corral on the west
bank of the Deschutes River; Horse-
thief Lake State Park; Horsethief Butte.
Evening-primrose Family.

374
Gayophytum diffusum　　　　I
Spreading Gayophytum　　(4-12 in.)
Dry open ground in forest openings at
middle to high elevations in the middle
Gorge.
MID JULY: old logging roads west of
the White Salmon River; near the top
of Monte Carlo.
Evening Primrose Family.

375
Oenothera contorta　　　　II
Slender Evening-primrose　(3-6 in.)
Dry sandy areas at low elevations east
from Crates Point.
MID MAY: Horsethief Butte; The
Dalles Bridge Road (US-197); Crates
Point Dunes.
Evening-primrose Family.

376
Oenothera pallida   III
Pale Evening-primrose   (6-24 in.)
Sandy areas, especially dunes, at low
elevations in the east Gorge.
LATE MAY-EARLY JUNE: along The
Dalles Bridge Road (US-197); dunes
near The Dalles Airport; Crates Point
Dunes.
Evening-primrose Family.

377
Oenothera villosa   IV
(Oenothera strigosa)
Common Evening-primrose   (1-4 ft.)
Fields and roadsides at low elevations
throughout the Gorge, but most
common in the west Gorge.
JULY: I-84 at Rooster Rock State Park.
Oenothera hookeri, with larger flow-
ers, is also found occasionally in the
west Gorge; blooms in July and August
on the banks of the Sandy River.
Evening-primrose Family.

GINSENG FAMILY

378
Oplopanax horridum   V
Devil's Club   (3-10 ft.)
Damp areas in coniferous forests in the
west Gorge.
EARLY MAY: Scenic Highway east of
Crown Point.
LATE MAY: Larch Mountain Highway.
Ginseng Family.

379
Angelica arguta   II
Shining Angelica   (2-5 ft.)
Meadows and forest openings at high
elevations in the west Gorge.
MID-LATE JULY: Silver Star Mountain;
Road 68 near Big Huckleberry
Mountain.
Parsley Family.

380
Angelica genuflexa   III
Kneeling Angelica   (3-8 ft.)
Margins of streams and waterfalls; wet
meadows at all elevations in the west
Gorge.
LATE JULY: Latourell Falls; Multnomah
Falls.
SEPT: Wahtum Lake.
Parsley Family.

381
Cicuta douglasii   II
Western Water-hemlock   (2-6 ft.)
Columbia River shores and bottom-
lands, and wet meadows at middle to
high elevations in the west and middle
Gorge.
MID AUG: Columbia River shore west
of Rooster Rock; Dalton Point; near
the mouth of Bridal Veil Creek; Sand
Island.
Parsley Family.

382
Cymopterus terebinthinus I
Indian-parsnip (6-24 in.)
Open rocky slopes near Grassy Knoll.
LATE MAY-EARLY JUNE: Road 68
below Grassy Knoll.
Parsley Family.

383*
Daucus carota V
Wild Carrot (1-3 ft.)
Roadsides and other disturbed areas,
generally at low elevations as far east
as The Dalles.
AUG: I-84 in the west Gorge; Scenic
Highway at Ainsworth State Park; SR-
14 from Dog Mountain to the White
Salmon River.
Parsley Family.

384
Heracleum lanatum V
Cow-parsnip (3-9 ft.)
Moist open woods and woodland
edges at all elevations in the west and
middle Gorge.
EARLY JUNE: I-84 east of the Sandy
River; Corbett Hill Road; Scenic High-
way from Crown Point to Ainsworth
State Park.
Parsley Family.

385
Ligusticum apiifolium   II
Parsley-leaf Lovage   (1.5-5 ft.)
Open woods at low to middle eleva-
tions as far east as Wind Mountain.
EARLY-MID JUNE: SR-14 near MP 20;
Gibson Road and other backroads
west of Cape Horn.
MID JUNE: Hamilton Mountain Trail.
Parsley Family.

386
Ligusticum grayi   II
Gray's Lovage   (1-2 ft.)
Meadows and open woods at higher
elevations in the west Gorge.
LATE JULY: near Indian Springs.
Parsley Family.

387
Lomatium canbyi   I
Canby's Desert Parsley   (2-6 in.)
Open slopes at all elevations east of
Lyle Tunnel; thus far found only on the
Washington side.
EARLY-MID MAR: near the microwave
station at the top of The Dalles Moun-
tain Road.
Parsley Family.

388
Lomatium columbianum IV
Columbia Desert Parsley (1-2.5 ft.)
Open slopes between the Little White
Salmon River and The Dalles.
MID MAR: SR-14 east of Lyle Tunnel
(MP 77-81); Old Highway at Major
Creek; I-84 east of the Memaloose
Rest Area.
MID MAY: near the top of Monte
Carlo.
Parsley Family.

389
Lomatium dissectum IV
var. dissectum
Fern-leaf Desert Parsley (2-6 ft.)
Open woods between Bridal Veil and
The Dalles, often near oaks.
LATE APRIL-EARLY JUNE: Multnomah
Falls; Dog Creek Falls; White Salmon
River Road; backroads between Hood
River and The Dalles.
The yellow form can be seen along SR-
14 about one mile east of the Maryhill
Museum.
Parsley Family.

390
Lomatium grayi V
Pungent Desert Parsley (6-24 in.)
Open rocky areas at low to middle
elevations, particularly between Dog
Mountain and The Dalles area.
EARLY APRIL: SR-14 east of Bingen
and at Tunnel #5; west side of the
Klickitat River canyon.
Parsley Family.

391
Lomatium laevigatum　II
Smooth Desert Parsley　(6-24 in.)
Open rocky areas and basalt cliffs at
low elevations along the Columbia
River east from The Dalles.
EARLY MAR: rocky shores of the
Columbia River near The Dalles Dam
Visitor Center.
MID MAR: cliffs at the east edge of
Wishram.
LATE MAR-EARLY APRIL: cliffs at the
north end of the US-97 bridge.
Parsley Family.

392
Lomatium leptocarpum　III
Slender-fruited　(6-20 in.)
Desert Parsley
Open slopes and woods at all eleva-
tions from Dog Mountain to the east
end of the Gorge.
MID APRIL: The Dalles Mountain
Road; SR-14 at DNR Falls.
EARLY MAY: top of Carroll Road; top
of The Dalles Mountain Road.
Parsley Family.

393
Lomatium macrocarpum　III
(yellow form)
Gray-leaf Desert Parsley,　(2-14 in.)
Biscuit Root
Dry open ground at all elevations
between Hood River and The Dalles.
EARLY APRIL: Old Highway near the
Lyle-Appleton Road junction; Klickitat
Fisherman's Park; Rock Creek Road.
MID-LATE APRIL: backroads between
Hood River and The Dalles; Hood
River Mountain Meadow.
Parsley Family.

394
Lomatium macrocarpum II
(white form)
Gray-leaf Desert Parsley,   (2-10 in.)
Biscuit Root
Dry open ground at all elevations east
of The Dalles.
LATE MAR-EARLY APRIL: Horsethief
Lake State Park.
MID-APRIL: top of The Dalles Moun-
tain Road.
Parsley Family.

395
Lomatium martindalei IV
Martindale's Desert   (3-14 in.)
Parsley
Open rocky areas at all elevations
between Crown Point and Mt. Defi-
ance.
EARLY MAY: Scenic Highway at
Shepperd's Dell, Multnomah Falls, and
Oneonta Gorge.
EARLY JUNE: top of Larch Mountain.
Parsley Family.

396
Lomatium nudicaule IV
Bare-stem Desert Parsley   (6-24 in.)
Open grassy slopes, fields, and road-
sides from Mt. Hamilton to the east
end of the Gorge, but mostly between
Hood River and The Dalles.
MID APRIL: SR-14 east of Lyle Tunnel
(MP 77-81); The Dalles Mountain
Road.
LATE APRIL-EARLY MAY: Old High-
way; backroads between Mosier and
The Dalles.
Parsley Family.

397
Lomatium piperi   IV
Piper's Desert Parsley,   (2-6 in.)
Salt-and-pepper
Open grassy areas at all elevations,
mostly between Hood River and The
Dalles.
EARLY MAR: Old Highway; SR-14
east of Lyle Tunnel (MP 77-81).
Parsley Family.

398
Lomatium suksdorfii   II
Suksdorf's Desert Parsley   (2-6 ft.)
Grasslands or open woods at middle to
high elevations between the Little
White Salmon River and The Dalles.
MID APRIL: lower and middle sections
of the Chenoweth Road.
LATE APRIL-EARLY MAY: upper
section of the Chenoweth Road;
Major Creek Road.
LATE MAY-EARLY JUNE: top of Nestor
Peak.
Parsley Family.

399
Lomatium triternatum   V
var. triternatum
Nine-leaf Desert Parsley   (6-30 in.)
Open areas at all elevations throughout
the Gorge.
LATE APRIL-EARLY MAY: SR-14 at
Dog Mountain and near Tunnel #5.
MID MAY: SR-14 at Cape Horn;
Beacon Rock.
Parsley Family.

400
Oenanthe sarmentosa   III
Water Parsley   (1-3 ft.)
Ditches, swamps, and margins of
streams and ponds, generally at low
elevations as far east as Mosier.
EARLY JULY: Latourell Falls; Shep-
perd's Dell; Bridal Veil Creek; Multno-
mah Falls; Home Valley Park; Franz
Lake.
Parsley Family.

401
Osmorhiza chilensis   IV
Common Sweet Cicely   (1-3 ft.)
Woodlands at all elevations as far east
as Mosier.
MID MAY: Scenic Highway at Bridal
Veil Creek and Ainsworth State Park;
I-84 at Starvation Creek State Park.
EARLY JUNE: Monte Carlo Trailhead.
Parsley Family.

402
Osmorhiza occidentalis   II
Mountain Sweet Cicely   (1.5-4 ft.)
Open woods in the west Gorge,
commonly at higher elevations.
LATE MAY: SR-14 just west of Cape
Horn; Mt. Zion Road.
LATE JUNE-EARLY JULY: Pacific Crest
Trail near Chinidere Mountain.
Parsley Family.

403
Perideridia gairdneri   II
Yampah   (1-4 ft.)
Meadows and grasslands at scattered
locations, especially in the east Gorge.
AUG: Tom McCall Nature Preserve.
LATE AUG: top of The Dalles Mountain
Road.
Similar but smaller *Perideridia oregana*
is found at scattered locations in the
west Gorge, blooming in late June-
early July.
Parsley Family.

404
Sanicula crassicaulis   II
Western Snake Root   (1-2 ft.)
Open woods, especially in the east
Gorge.
MID MAY: woods along SR-14 near
MP 20; oak woods near the Klickitat
Fish Ladder.
Parsley Family.

405
Sanicula graveolens   III
Sierra Snake Root   (3-12 in.)
Open places and lightly wooded areas,
especially between Dog Mountain and
Lyle, often at higher elevations.
LATE APRIL: oak woods near Catherine
Creek; Major Creek Road.
EARLY MAY: Grassy Knoll Trailhead.
Parsley Family.

406
Tauschia stricklandii   I
Strickland's Tauschia   (4-10 in.)
Wet subalpine meadows just west of
Tanner Creek at 3,500 ft. elev.
LATE JUNE-EARLY JULY: pond along
Moffett Trail near the rim of Tanner
Creek Valley.
Parsley Family.

--------------------------------------- DOGWOOD FAMILY

407
Cornus canadensis   IV
Canadian Dogwood,   (4-8 in.)
Bunchberry
Woodlands at middle to high elevations
in the west Gorge.
MID JUNE: Larch Mountain Corridor.
EARLY JULY: top of Larch Mountain.
Dogwood Family.

408
Cornus nuttallii   II
Pacific Dogwood   (10-50 ft.)
Woodlands at low to middle elevations
as far east as the Mosier area.
LATE APRIL-EARLY MAY: I-84 at
Rooster Rock State Park, Cascade
Locks, and Hood River. Trees bloom in
late summer as well as spring on SR-14
at Tunnels #2 and #4.
Dogwood Family.

409
Cornus stolonifera   III
var. occidentalis
Creek Dogwood   (6-18 ft.)
Wet areas at low elevations throughout
the Gorge.
MID-LATE MAY: I-84 near the Sandy
River, at Rooster Rock State Park, and
at Dalton Point; Scenic Highway near
Crown Point and at Ainsworth State
Park; SR-14 at Horsethief Butte.
Dogwood Family.

HEATH FAMILY

410
Allotropa virgata   I
Candy Stick   (5-15 in.)
Coniferous woods at middle to high
elevations in the west Gorge.
LATE JUNE-EARLY JULY: South Prairie
Road at MP 5-6.
Heath Family.

411
Arctostaphylos   III
columbiana
Manzanita   (3-8 ft.)
Dry woodlands in the west Gorge,
mostly between Bonneville Dam and
the White Salmon River.
MID MAR: I-84 from Starvation Creek
State Park to Hood River; SR-14 at
Wind Mountain.
Heath Family.

412
Arctostaphylos uva-ursi III
Kinnikinnick (4-8 in.)
Dry open places at either low or high
elevations as far east as the Major
Creek Plateau.
LATE MAR-EARLY APRIL: near the old
navigation lock at Bonneville Dam;
abandoned quarry at Government
Cove.
LATE APRIL: bluff at Bridal Veil Falls
State Park.
Heath Family.

413
Chimaphila menziesii II
Little Pipsissewa (3-6 in.)
Coniferous woods at middle to high
elevations in the west and middle
Gorge.
LATE JUNE-EARLY JULY: woods on
the Major Creek Plateau; South Prarie
Road at MP 5-6.
LATE JULY-EARLY AUG: Spring Camp
Road.
Heath Family.

414
Chimaphila umbellata II
Pipsissewa (6-12 in.)
Coniferous woods, mostly at middle to
high elevations in the west Gorge.
MID JULY: South Prairie Road at MP 5.
EARLY AUG: Rainy Lake Road near
Mt. Defiance; Warren Lake Trail.
Heath Family.

415
Gaultheria ovatifolia   II
Slender Wintergreen   (1-2 in.)
Coniferous woods and the edges of
marshes at middle to high elevations in
the west Gorge.
EARLY JULY: stream at the North Lake
Trailhead.
Heath Family.

416
Gaultheria shallon   IV
Salal   (1-3 ft.)
Coniferous forests at low to middle
elevations in the west Gorge.
LATE MAY: roads at Home Valley.
LATE JUNE-EARLY JULY: Larch Moun-
tain Corridor; Spring Camp Road.
Heath Family.

417
Hemitomes congestum   I
Gnome Plant   (1-5 in.)
Woods at higher elevations in the
middle Gorge.
LATE JULY-EARLY AUG: thus far
found in the Gorge only east of Mt.
Defiance in dry Douglas-fir woods at
3,100 ft. elev.
Heath Family.

418
Hypopitys monotropa II
Pinesap (2-10 in.)
Deep shady woods at middle to high
elevations in the west Gorge.
LATE JULY: top of Larch Mountain.
Heath Family.

419
Kalmia occidentalis II
Western Swamp Laurel (6-18 in.)
Middle to high elevation marshes in
the west Gorge.
MID-LATE MAY: marshes at the head
of Horsetail Creek.
Heath Family.

420
Menziesia ferruginea II
var. ferruginea
Fool's Huckleberry (3-6 ft.)
Margins of lakes and marshes at
middle to high elevations in the west
Gorge.
LATE JUNE: Indian Springs Road.
EARLY JULY: North Lake Trailhead;
Rainy Lake.
Heath Family.

421
Monotropa uniflora   II
Indian Pipe   (2-10 in.)
Coniferous woods at low to middle
elevations in the west and middle
Gorge.
LATE JUNE: Fort Cascades Historic
Site.
EARLY JULY: McCord Creek Falls Trail.
Heath Family.

422
Phyllodoce empetriformis   I
Red Mountain Heather   (6-18 in.)
Moist open areas at scattered high
elevation sites in the west Gorge.
EARLY JULY: near Indian Springs.
Heath Family.

423
Pterospora andromedea   II
Pinedrops   (1-3 ft.)
Coniferous woods at low to middle
elevations in the west and middle
Gorge.
LATE JULY: South Prairie Road at MP
4-5; backroads on the Major Creek
Plateau.
Heath Family.

424
Pyrola asarifolia II
Large Pyrola (6-16 in.)
Coniferous woods at middle to high
elevations in the west Gorge.
EARLY-MID JULY: Warren Lake
Trailhead.
MID JULY: Rainy Lake.
LATE JULY-EARLY AUG: Pacific Crest
Trail between Wahtum Lake and
Indian Mountain.
Heath Family.

425
Pyrola picta II
White-vein Pyrola (6-12 in.)
Coniferous woods at middle to high
elevations between Bridal Veil and
Lyle.
LATE JUNE-EARLY JULY: backroads
east and west of Augspurger Moun-
tain; backroads on the Major Creek
Plateau; South Prairie Road near MP 5.
Heath Family.

426
Pyrola secunda III
var. secunda
Sidebells Pyrola (3-6 in.)
Coniferous woods at higher elevations
in the west Gorge.
LATE JULY: North Lake Trailhead.
AUG: primitive road (gated) north of
Wahtum Lake Forest Camp.
Heath Family.

427
Rhododendron   II
macrophyllum
Western Rhododendron   (3-15 ft.)
Woods at middle to high elevations in
the west Gorge.
LATE JUNE: Spring Camp Road; primi-
tive road (gated) north of Wahtum
Lake Forest Camp.
Heath Family.

428
Vaccinium   III
membranaceum
Black Huckleberry   (2-6 ft.)
Above 2,000 ft. elev. in the west
Gorge.
MID-LATE JUNE: Pacific Crest Trail
near Big Huckleberry Mountain; top of
Larch Mountain. Several other species
of huckleberry are also found in the
Gorge.
Heath Family.

PRIMROSE FAMILY

429
Dodecatheon conjugens   III
Desert Shooting Star   (4-12 in.)
Open places at all elevations from
Rowland Lake to the east end of the
Gorge.
LATE MAR-EARLY APRIL: grassy areas
along Klickitat River Road; Bobs Point;
benches above the Columbia River
east of The Dalles.
Primrose Family.

430
Dodecatheon cusickii   I
Sticky Shooting Star   (6-14 in.)
Open grassy slopes at low elevations
east of Wishram.
EARLY APRIL: flats overlooking the
Columbia River west of Biggs Junction.
Primrose Family.

431
Dodecatheon dentatum   III
White Shooting Star   (5-16 in.)
Waterfalls and wet cliffs as far east as
Starvation Creek State Park, mostly at
low elevations.
LATE MAY: Latourell Falls; Multnomah
Falls; Starvation Creek Falls.
Primrose Family.

432
Dodecatheon hendersonii   I
Oval-leaf Shooting Star   (4-12 in.)
A Willamette Valley species that grows
on flat, grassy areas near oaks and
brush.
MID-LATE APRIL: thus far found in the
Gorge only on private land on SR-14 at
MP 62.
Primrose Family.

433
Dodecatheon jeffreyi   III
Tall Mountain   (8-20 in.)
Shooting Star
Marshes and springs at middle to high
elevations in the west Gorge.
LATE JUNE: primitive road (gated)
north of Wahtum Lake Forest Camp;
North Lake Trailhead; Rainy Lake.
Primrose Family.

434
Dodecatheon poeticum   IV
Poet's Shooting Star   (6-14 in.)
Open or lightly wooded moist slopes
and streamsides between Dog Moun-
tain and Horsethief Butte.
LATE MAR-EARLY APRIL: Memaloose
Viewpoint; Tom McCall Nature Pre-
serve; SR-14 east of Lyle Tunnel (MP
77-81).
MID APRIL: Dry Creek Road at MP
3-4.
Primrose Family.

435
Dodecatheon pulchellum   II
Few-flowered   (6-12 in.)
Shooting Star
Waterfalls, wet cliffs, seeps, and
streamsides in the west Gorge; thus far
found only on the Oregon side.
LATE APRIL-EARLY MAY: Gorge Trail
at waterfalls approximately one mile
east of Multnomah Falls.
Primrose Family.

436
Douglasia laevigata II
var. laevigata
Smooth-leaf Douglasia (2-6 in.)
Basalt cliffs and rock outcrops at low to middle elevations on both sides of the Columbia River between Crown Point and Mitchell Point.
LATE MAR-EARLY APRIL: I-84 at Mitchell Point.
EARLY APRIL: trail to the top of McCord Creek Falls.
Primrose Family.

437
Lysimachia ciliata III
Fringed Loosestrife (1-3 ft.)
Damp bottomlands along the Columbia River in the west Gorge.
LATE JULY-EARLY AUG: Mirror Lake; Franz Lake.
Primrose Family.

438*
Lysimachia terrestris I
Bog Loosestrife, (1-2 ft.)
Swamp Candle
Species native to eastern section of North America.
MID JULY: thus far found in the Gorge only along the north shore of Ashes Lake.
Primrose Family.

439
Trientalis arctica   II
Northern Star Flower   (3-6 in.)
Wet meadows and marshy shores of
lakes at higher elevations in the west
Gorge.
LATE JUNE-EARLY JULY: spring at the
North Lake Trailhead; Rainy Lake.
Primrose Family.

440
Trientalis latifolia   IV
Broad-leaf Star Flower   (4-8 in.)
Open coniferous woods at all eleva-
tions in the west and middle Gorge.
LATE MAY: Starvation Creek Falls;
Beacon Rock.
Primrose Family.

GENTIAN FAMILY

441*
Centaurium erythraea   II
Common Centaury   (6-18 in.)
Moist open ground at low elevations
throughout the Gorge.
MID JULY: Rooster Rock State Park;
SR-14 at the Clark-Skamania County
Line; Grant Lake.
Gentian Family.

442
Centaurium exaltatum I
Western Centaury (4-10 in.)
SUMMER: shores of low elevation
ponds east of The Dalles.
Gentian Family.

443
Frasera albicaulis III
var. columbiana
Columbia Frasera (1-2 ft.)
Dry open ground at all elevations in
the east Gorge.
LATE MAY-EARLY JUNE: Tom McCall
Nature Preserve south of US-30.
MID JUNE: The Dalles Mountain
Road; Hood River Mountain Meadow.
Gentian Family.

444
Gentiana calycosa II
Explorer's Gentian (8-20 in.)
Rocky forest openings at high eleva-
tions in the west Gorge.
LATE AUG-EARLY SEPT: Indian Moun-
tain; Silver Star Mountain.
Gentian Family.

445
Gentiana sceptrum   II
Staff Gentian   (1-2 ft.)
EARLY AUG: marshes at higher eleva-
tions west of Bonneville Dam.
Gentian Family.

BUCKBEAN FAMILY

446
Menyanthes trifoliata   II
Buckbean   (6-18 in.)
LATE MAY-EARLY JUNE: marshes and
margins of ponds at middle to high
elevations in the west Gorge.
Buckbean Family.

DOGBANE FAMILY

447
Apocynum   IV
androsaemifolium
Flytrap Dogbane   (6-16 in.)
Dry, open or lightly wooded areas and
roadsides. Most common between
Dog Mountain and The Dalles.
LATE JUNE-EARLY JULY: backroads
between Mosier and The Dalles;
Snowden Road out of White Salmon.
Dogbane Family.

448
Apocynum cannabinum II
Hemp Dogbane (1-3 ft.)
Columbia River shores throughout the
Gorge. Blooming times are dependent
on the level of the river.
MID AUG: Dalton Point; east bank of
the Deschutes River near Hwy 206.
Closely related *Apocynum sibiricum* is
also found in the Gorge.
Dogbane Family.

449
Asclepias fascicularis III
Narrow-leaf Milkweed (1-2.5 ft.)
Open areas at low elevations in the
east Gorge, often in ditches and the
margins of vernal ponds.
LATE JUNE: The Dalles Bridge Ponds;
SR-14 at the junction with The Dalles
Bridge Road (US-197).
LATE JULY: Stevens Pond.
Milkweed Family.

450
Asclepias speciosa I
Showy Milkweed (1.5-4 ft.)
Ditches and moist open ground at low
elevations from Mosier to the east end
of the Gorge.
LATE JUNE: SR-14 at MP 91-93; Celilo
Park; west bank of the Deschutes
River; US-30 east of Mosier (MP 1.7).
Milkweed Family.

451*
Convolvulus arvensis   III
Field Morning Glory   (vine)
Fields and waste ground at low to
middle elevations throughout the
Gorge.
LATE JUNE: Grant Lake; The Dalles
Mountain Road; SR-14 near the The
Dalles Bridge Road (US-197) junction.
The introduced *Convolvulus sepium* is
also found at low elevations through-
out the Gorge.
Morning Glory Family.

452
Convolvulus nyctagineus   I
Night-blooming   (8-18 in.)
Morning Glory
Open, lighly wooded, or brushy slopes
at middle to high elevations between
Dog Mountain and the White Salmon
River.
MID JULY: Monte Carlo Trail.
Morning Glory Family.

453
Collomia grandiflora   III
Large-flowered Collomia   (1-3 ft.)
Dry, open or lightly wooded areas,
mainly between Hood River and The
Dalles.
LATE MAY-EARLY JUNE: SR-14 east of
Lyle Tunnel (MP 77-81); backroads
between Mosier and The Dalles.
EARLY JULY: Scenic Highway at MP
10.2.
Phlox Family.

454
Collomia heterophylla   III
Varied-leaf Collomia   (3-12 in.)
Open woods at low to middle eleva-
tions, mostly in the west Gorge.
MID MAY: Dog Mountain Trail.
EARLY JUNE: Buck Creek Road; Hamil-
ton Mountain Trail; shoulders of the
Scenic Highway at Ainsworth State
Park.
Phlox Family.

455
Gilia capitata   V
Blue Field Gilia   (6-36 in.)
Dry open ground at low to middle
elevations, mostly in the west Gorge.
MID JUNE: SR-14 at Wind Mountain
and Dog Mountain.
EARLY JULY: Scenic Highway at Crown
Point and Ainsworth State Park.
MID JULY: SR-14 at Cape Horn.
Phlox Family.

456
Gilia sinuata   I
Shy Gilia   (4-12 in.)
Sandy soil in the east Gorge.
LATE APRIL: edge of dunes on Miller
Island.
Phlox Family.

457
Linanthus bakeri   I
Baker's Linanthus   (2-6 in.)
Barren, generally south-facing slopes
in the east Gorge.
MID-LATE APRIL: thus far found in the
Gorge only in the Mosier Creek drain-
age.
Phlox Family.

458
Linanthus bicolor   IV
Baby Stars   (2-6 in.)
Open grasslands between Hood River
and Lyle, mostly at low elevations.
EARLY MAY: Old Highway near
Catherine Creek; Memaloose View-
point; Tom McCall Nature Preserve.
LATE MAY-EARLY JUNE: Hood River
Mountain Meadow.
Phlox Family.

459
Microsteris gracilis   V
Midget Phlox   (3-8 in.)
Open grassy areas at all elevations
from Hamilton Mountain to the east
end of the Gorge.
LATE MAR-EARLY APRIL: Dog Creek
Falls; Old Highway; SR-14 east of Lyle
Tunnel (MP 77-81).
EARLY APRIL: Tom McCall Nature
Preserve.
Phlox Family.

460
Navarretia divaricata   I
Mountain Navarretia   (1-2 in.)
Dried mud of roads and ditches at
middle to high elevations in the middle
Gorge. Thus far found only on the
Washington side.
EARLY-MID JULY: top of Monte Carlo.
Phlox Family.

461
Navarretia intertexta   III
Needle-leaf Navarretia   (2-8 in.)
Moist ground, particularly in the
drying soil of ditches, vernal ponds,
and intermittent streams, at all elevations between Bingen and The Dalles
area.
MID JUNE: Tom McCall Nature Preserve; vernal ponds in Horsethief Lake
State Park; dry streambeds above SR-
14 at MP 70.4.
Phlox Family.

462
Navarretia squarrosa   II
Skunkweed   (4-10 in.)
Fields, roadsides, and other disturbed
areas as far east as The Dalles.
LATE JUNE-EARLY JULY: SR-14 at the
Clark-Skamania County Line; roadsides
on the Major Creek Plateau.
Phlox Family.

463
Navarretia tagetina   I
Marigold Navarretia   (4-7 in.)
MID JUNE: thus far found in the Gorge
only in vernally moist open ground
and ditches along the Old Highway
near Catherine Creek.
Phlox Family.

464
Phlox diffusa   III
Spreading Phlox   (2-6 in.)
Rocky slopes at all elevations in the
west Gorge.
LATE APRIL: SR-14 near Dog Creek
Falls and near Cook.
MID MAY: Hamilton Mountain Trail.
LATE MAY: Grassy Knoll Trailhead.
Phlox Family.

465
Phlox hoodii   I
Hood's Phlox   (2-4 in.)
Broad open ridge of the Columbia
Hills.
MID APRIL: top of The Dalles Moun-
tain Road.
LATE APRIL: gate on Haystack Butte
Road.
Phlox Family.

466
Phlox longifolia ‖
Long-leaf Phlox (4-16 in.)
Dry, open sandy ground at mostly low elevations east of The Dalles.
MID APRIL: BLM Parcel at Celilo; SR-14 near MP 91; near Avery Gravel Pit.
EARLY-MID MAY: The Dalles Bridge Road (US-197); Bobs Point; I-84 east of The Dalles near the turnout at MP 92.3.
Phlox Family.

467
Phlox speciosa ‖
Showy Phlox (4-16 in.)
Open woods and grasslands at middle to high elevations from Viento State Park to the east end of the Gorge.
MID APRIL: halfway up The Dalles Mountain Road.
MID MAY: backroads between Mosier and The Dalles; top of The Dalles Mountain Road.
Phlox Family.

468
Polemonium carneum ‖
Salmon Polemonium (1-3 ft.)
Brushy areas and forest openings at middle elevations in a narrow range of the west Gorge.
MID-LATE JUNE: old logging roads near Prindle Mountain; woods near the junction of Mabee Mines Road and McCloskey Creek Road; near Devils Rest.
Phlox Family.

SECTION FOUR

469
Hesperochiron pumilus   I
Dwarf Hesperochiron   (2-4 in.)
Moist swales in The Dalles area.
EARLY-MID MAY: swales south of The
Dalles along Skyline Road at Dutch
Flat.
Waterleaf Family.

470
**Hydrophyllum capitatum   III
var. thompsonii**
Ball-head Waterleaf   (4-16 in.)
Open or lightly wooded slopes at all
elevations between Dog Mountain
and The Dalles.
LATE MAR-EARLY APRIL: SR-14 at
MP 77.3; US-30 at Crates Point.
MID APRIL: Memaloose Rest Area;
upper Chenoweth Road.
LATE APRIL: Hood River Mountain
Meadow.
Waterleaf Family.

471
Hydrophyllum tenuipes   IV
Pacific Waterleaf   (1-2.5 ft.)
Moist woods at low to middle eleva-
tions as far east as Dog Mountain.
MID MAY: Latourell Falls; Ainsworth
State Park.
Waterleaf Family.

472
Nemophila parviflora IV
Woods Nemophila (2-8 in.)
Open woods at low to middle eleva-
tions in the west and middle Gorge.
LATE APRIL-EARLY MAY: base of
Rooster Rock; Dog Creek Falls; Beacon
Rock Trail; Horsetail Falls Trail.
Waterleaf Family.

473
Nemophila pedunculata II
Meadow Nemophila (1-3 in.)
Open grassy areas at low to middle
elevations between Bingen and The
Dalles area.
LATE MAR-EARLY APRIL: Memaloose
Rest Area; Tom McCall Nature Pre-
serve; SR-14 east of Lyle Tunnel (MP
77-81).
Waterleaf Family.

474
Phacelia hastata IV
Silver-leaf Phacelia (6-20 in.)
Dry open places at all elevations east
of Wind Mountain, often in roadsides
and other disturbed areas.
LATE MAY: Dog Creek Falls; SR-14
east of Bingen; Memaloose Rest Area;
The Dalles Bridge Ponds; top of The
Dalles Mountain Road.
Waterleaf Family.

475
Phacelia heterophylla   II
Varied-leaf Phacelia   (1-3 ft.)
Dry open places at low elevations in
the east Gorge, but also middle to high
elevations in the west and middle
Gorge.
JULY: near Indian Springs.
Waterleaf Family.

476
Phacelia linearis   III
Thread-leaf Phacelia   (2-20 in.)
Dry open places at low to middle
elevations, generally east of Dog
Mountain.
MID APRIL: SR-14 east of Lyle Tunnel
(MP 77-81).
LATE APRIL: sandy places along The
Dalles Bridge Road (US-197).
LATE MAY: Dog Creek Falls.
Waterleaf Family.

477
Phacelia nemoralis   III
Woodland Phacelia   (2-6 ft.)
Open woods and roadsides at low to
middle elevations in the west Gorge.
JULY: Scenic Highway at Crown Point,
Latourell Falls, and Wahkeena Falls.
Waterleaf Family.

478
Phacelia ramosissima   I
Branching Phacelia   (1-2 ft.)
Talus slopes and basalt ledges at low
elevations east of The Dalles.
LATE MAY: DNR Falls.
Waterleaf Family.

479
Romanzoffia sitchensis   III
Sitka Mist Maidens   (4-8 in.)
Mossy banks and cliffs at low eleva-
tions in the west Gorge, mostly on the
Oregon side.
LATE MAR-EARLY APRIL: Multnomah
Falls; Oneonta Gorge; Lower Tanner
Creek Road.
LATE APRIL: Corbett Hill Road.
Waterleaf Family.

BORAGE FAMILY

480
Amsinckia lycopsoides   III
Tarweed Fiddleneck   (6-36 in.)
Roadsides and fields in the east Gorge,
typically at low elevations.
APRIL: SR-14 east of Lyle Tunnel (MP
77-81) and at Horsethief Butte; BLM
Parcel at Celilo.
LATE APRIL: I-84 at Mayer State Park
(MP 75).
Borage Family.

481
Amsinckia menziesii I
Small-flowered Fiddleneck (1-2 ft.)
Scattered locations as far east as The
Dalles.
MID MAY: near Oneonta Bridge;
Columbia River shore at Cape Horn.
LATE MAY-EARLY JUNE: meadow at
DNR Forest.
Borage Family.

482
Amsinckia retrorsa IV
Rigid Fiddleneck (6-36 in.)
Open grasslands in the east Gorge.
APRIL: Old Highway; SR-14 east of
Lyle Tunnel (MP 77-81).
MAY: top of The Dalles Mountain
Road.
Borage Family.

483
Cryptantha celosioides I
Cockscomb Cryptantha (6-16 in.)
Dry slopes in the east Gorge.
LATE APRIL-EARLY MAY: SR-14 near
MP 98.4.
MID MAY: top of Haystack Butte.
Borage Family.

484
Cryptantha flaccida   IV
Weak-stem Cryptantha   (4-16 in.)
Dry open areas at low to middle elevations in the east Gorge.
LATE APRIL-EARLY MAY: Old Highway; SR-14 at MP 70.4 and east of Lyle Tunnel (MP 77-81).
Borage Family.

485
Cryptantha intermedia   V
Common Cryptantha   (4-16 in.)
Dry, open or lightly wooded slopes as far east as Wishram.
LATE APRIL-EARLY MAY: SR-14 at Wind Mountain, Dog Creek Falls, and east of Lyle Tunnel (MP 77-81).
JUNE: Scenic Highway.
JULY: SR-14 at Cape Horn.
Borage Family.

486
Cryptantha rostellata   I
Red-stem Cryptantha   (6-12 in.)
Barren, generally south-facing slopes.
LATE APRIL-EARLY MAY: thus far found in the Gorge only in the Mosier Creek drainage.
Borage Family.

487
Cynoglossum grande   III
Great Hound's Tongue   (1-2.5 ft.)
Oak woods at low to middle elevations
between Grant Lake and the Klickitat
River.
EARLY APRIL: Grant Lake; woods
along SR-14 just west of Bingen;
Klickitat Fish Ladder.
MID APRIL: Rock Creek Road at MP
0.6 (watch for ticks here).
Borage Family.

488
Hackelia diffusa   II
var. cottonii
Branching Stickseed   (6-24 in.)
(white-flowered variety)
Shaded cliffs and talus slopes in the
east Gorge, chiefly on the Oregon
side.
MID MAY: cliffs along I-84 between
Hood River and Biggs Junction.
LATE MAY: low cliffs on US-30 at the
west edge of Mayer State Park (MP
6.5).
Borage Family.

489
Hackelia diffusa   I
var. diffusa
Branching Stickseed   (1-2 ft.)
(blue-flowered variety)
Shaded cliffs and talus slopes in the
west Gorge, commonly on the Oregon
side.
LATE MAY-EARLY JUNE: McCord
Creek Falls; Oneonta Gorge.
Borage Family.

490
Lithospermum ruderale   III
Stoneseed, Puccoon,   (1-2 ft.)
Gromwell
Dry open fields and slopes at all elevations from Dog Mountain to the east end of the Gorge.
MID APRIL: Horsethief Butte; The Dalles Mountain Road; BLM Parcel at Celilo.
Borage Family.

491
Mertensia oblongifolia   I
Prairie Mertensia   (4-12 in.)
Grassy areas at higher elevations in the east Gorge.
LATE APRIL: thus far found in the Gorge only on top of the Columbia Hills in Washington.
Borage Family.

492
Mertensia paniculata   I
Tall Mountain Mertensia   (2-5 ft.)
Streambanks, wet cliffs, and talus slopes at higher elevations east of the crest of the Cascade Mountains.
MID-LATE JULY: thus far found in the Gorge only on talus slopes above North Lake.
Borage Family.

493
Myosotis laxa   III
Small-flowered   (2-12 in.)
Forget-me-not
Low elevation wet places, especially
Columbia River bottomlands through-
out the Gorge.
AUG: Mirror Lake; Rock Creek Lake;
Horsethief Lake.
Borage Family.

494 *
Myosotis sylvatica   II
Woods Forget-me-not   (4-16 in.)
MID MAY: Scenic Highway between
Crown Point and Ainsworth State
Park, especially in the vicinity of Bridal
Veil.
Borage Family.

495
Plagiobothrys figuratus   II
Fragrant Popcorn Flower   (3-10 in.)
Open areas which are wet in spring
and dry in summer, mainly between
Mosier and Lyle.
MID MAY: Old Highway near the
Lyle-Appleton Road junction; Tom
McCall Nature Preserve.
Borage Family.

496
Plagiobothrys   IV
nothofulvus
Rusty Popcorn Flower   (6-16 in.)
Open slopes at low to middle eleva-
tions between Dog Mountain and The
Dalles.
MID APRIL: SR-14 east of Lyle Tunnel
(MP 77-81); Lyle-Appleton Road.
Borage Family.

497
Plagiobothrys scouleri   II
Scouler's Popcorn Flower   (1-8 in.)
Vernal ponds, ditches, swales, and lake
margins in the east Gorge.
LATE MAY: Tom McCall Nature Pre-
serve.
Borage Family.

498
Plagiobothrys tenellus   V
Slender Popcorn Flower   (2-10 in.)
Dry open ground at all elevations in
the east Gorge.
MID APRIL: SR-14 east of Lyle Tunnel
(MP 77-81); Horsethief Butte; Bobs
Point.
LATE APRIL-EARLY MAY: Hood River
Mountain Meadow.
Borage Family.

499
Verbena bracteata   II
Bracted Verbena   (2-6 in.)
Roadsides, disturbed ground, and dry
rocky areas, chiefly in the east Gorge
at low elevations.
JULY: Rowland Lake; Horsethief Butte;
Bobs Point.
Verbena Family.

500
Verbena hastata   I
Blue Verbena   (1-3 ft.)
Columbia River bottomlands in the
vicinity of Bonneville Dam.
AUG: scattered populations on the
shores of Bradford Island, Ashes Lake,
and Rock Creek Lake.
Verbena Family.

MINT FAMILY

501*
Lamium purpureum   II
Red Dead-nettle   (4-8 in.)
Roadsides and other disturbed ground
in the west Gorge.
APRIL: Bridal Veil Falls State Park;
Ainsworth State Park.
Closely related *Lamium amplexicaule*
is found in similar sites in the east
Gorge.
Mint Family.

502
Lycopus americanus   III
Cut-leaf Bugleweed   (1-2 ft.)
Columbia River bottomlands through-
out the Gorge.
AUG: Mirror Lake; Rock Creek Lake;
Horsethief Lake; pond on I-84 at MP
94.4.
Mint Family.

503
Lycopus asper   I
Rough Bugleweed   (1-2 ft.)
Columbia River shores and bottom-
lands in The Dalles area and at scat-
tered locations to the east.
AUG: shore east of The Dalles Riverside
Park.
Mint Family.

504
Lycopus uniflorus   II
Northern Bugleweed   (8-16 in.)
Columbia River bottomlands and
upland ponds in the west Gorge.
LATE JULY-EARLY AUG: Young Creek
in Rooster Rock State Park; Home
Valley Park; Beacon Rock Pond.
Mint Family.

505
Mentha arvensis IV
Field Mint (1-2.5 ft.)
Columbia River shores and bottom-
lands throughout the Gorge; vernal
ponds in the east Gorge.
AUG: Dalton Point; Mirror Lake;
Beacon Rock shore; Home Valley Park.
Mint Family.

506
Monardella odoratissima I
Western Balm (6-24 in.)
Rocky areas at either low or high
elevations in the middle Gorge.
MID JULY: SR-14 at Tunnel #5.
MID AUG: Tomlike Mountain.
Mint Family.

507 *
Nepeta cataria II
Catnip (1-3 ft.)
Roadsides and streambanks at low
elevations thoughout the Gorge.
MID JULY: SR-14 at Lawton Creek;
Horsethief Butte; The Dalles Mountain
Road.
Mint Family.

508
Physostegia parviflora   I
Purple Dragon Head   (8-20 in.)
Columbia River bottomlands west of
Bonneville Dam.
LATE JULY: Young Creek in Rooster
Rock State Park; Franz Lake.
Mint Family.

509
Prunella vulgaris   IV
Self-heal   (4-16 in.)
Open woods and disturbed areas at
low to middle elevations in the west
and middle Gorge.
MID JUNE: Grant Lake; Bridal Veil Falls
State Park.
JULY: Lower Tanner Creek Road; SR-
14 at the Clark-Skamania County Line.
Mint Family.

510
Satureja douglasii   II
Yerba Buena   (1-6 in.)
Dry open coniferous woods at low to
middle elevations between Bonneville
Dam and The Dalles.
LATE JUNE: woods near Northwestern
Lake on the White Salmon River;
Gorge Trail at Ruckel Creek; Fort
Cascades Historic Site.
Mint Family.

511
Scutellaria angustifolia   II
Narrow-leaf Skullcap   (6-12 in.)
Dry open rocky places at low elevations
in the east Gorge; thus far found only
on the Washington side.
LATE MAY: SR-14 east of Lyle Tunnel
(MP 77-81); Horsethief Butte; DNR
Falls.
Mint Family.

512
Scutellaria lateriflora   II
Mad-dog Skullcap   (1-2.5 ft.)
Columbia River bottomlands, chiefly
west of Bonneville Dam.
AUG: Mirror Lake; Sand Island; Beacon
Rock Pond.
Mint Family.

513
Stachys cooleyae   IV
Great Hedge-nettle   (2-4 ft.)
Wet places such as lake and stream
margins, waterfalls, wet meadows,
and ditches at all elevations in the west
and middle Gorge.
LATE JULY-EARLY AUG: Latourell
Falls; Grant Lake; Dog Creek Falls.
Mint Family.

514
Stachys mexicana   I
Coast Hedge-nettle   (1-2 ft.)
Moist woods west of Angels Rest.
EARLY JULY: Angels Rest Trail; Scenic
Highway east of Crown Point near MP
11.8 (for safety, walk in the ditch).
Mint Family.

515
Stachys palustris var. pilosa   I
Swamp Hedge-nettle   (8-24 in.)
Columbia River shores and bottom-
lands in the middle and east Gorge.
LATE JULY-EARLY AUG: shore east of
The Dalles Riverside Park; shore near
Wyeth.
Mint Family.

POTATO FAMILY

516
Datura stramonium   I
Jimson Weed   (1-3 ft.)
LATE AUG-EARLY SEPT: found chiefly
along the banks of the Deschutes River
and occasionally along Columbia River
shores.
Potato Family.

517*
Nicotiana acuminata II
Wild Tobacco (1-3 ft.)
SEPT: sandy ground along the west
bank of the Deschutes River.
Potato Family.

518*
Solanum dulcamara III
Bittersweet Nightshade (2-6 ft.)
Moist thickets, Columbia River bottom-
lands, and open woods at low eleva-
tions throughout the Gorge.
LATE JULY-EARLY AUG: mouth of
Bridal Veil Creek; Sand Island;
Ainsworth State Park; Home Valley
Park.
Potato Family.

519
Solanum nigrum II
Black Nightshade (6-16 in.)
Dry sandy shores of the Columbia
River throughout the Gorge.
EARLY-MID SEPT: Beacon Rock shore;
west bank of the Deschutes River.
Potato Family.

520
Castilleja hispida IV
Harsh Paintbrush (6-24 in.)
Open slopes and rocky places through-
out the Gorge.
MID MAY: Scenic Highway at Shep-
perd's Dell; SR-14 at Cape Horn and at
the Clark-Skamania County Line.
Figwort Family.

521
Castilleja miniata IV
Common Paintbrush (6-30 in.)
Meadows and open woods as far east
as Lyle, generally at middle to high
elevations.
EARLY JUNE: Scenic Highway near the
Mt. Hood National Forest boundary
and near Multnomah Falls.
EARLY JULY: top of Larch Mountain.
Figwort Family.

522
Castilleja rupicola I
Cliff Paintbrush (6-10 in.)
Vertical basalt cliffs at low to middle
elevations in the west Gorge.
LATE APRIL-EARLY MAY: McCord
Creek Falls Trail.
Figwort Family.

523
Castilleja suksdorfii   I
Bog Paintbrush   (1-2 ft.)
JULY: marshes at higher elevations in
the west Gorge.
Figwort Family.

524
Collinsia grandiflora   IV
Large-flowered   (4-12 in.)
Blue-eyed Mary
Open slopes, rocky banks, and mossy
cliffs, mostly at low elevations in the
west Gorge.
LATE MAY: Scenic Highway near
Crown Point, Latourell Falls, and
Shepperd's Dell; Lower Tanner Creek
Road; SR-14 at the Clark-Skamania
County Line and at Signal Rock.
Figwort Family.

525
Collinsia parviflora   V
Small-flowered   (2-12 in.)
Blue-eyed Mary
Open areas and forest openings
throughout the Gorge.
EARLY APRIL: SR-14 east of Lyle
Tunnel (MP 77-81); Old Highway at
Catherine Creek.
MID APRIL: base of Rooster Rock;
Memaloose Viewpoint.
Figwort Family.

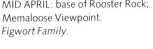

526
Collinsia rattanii II
Rattan's Blue-eyed Mary (4-12 in.)
Open woods from the White Salmon
River to The Dalles, usually at middle
elevations.
MID MAY: about three miles north of
Lyle on the Lyle-Appleton Road.
LATE MAY: Hood River Mountain
Meadow.
Figwort Family.

527
Collinsia sparsiflora II
Few-flowered (2-8 in.)
Blue-eyed Mary
Moist open slopes between Bingen
and The Dalles.
MID APRIL: Old Highway near Cather-
ine Creek and near the gravel pit just
east of Major Creek; Tom McCall
Nature Preserve.
LATE APRIL: primitive upper section of
Rock Creek Road; Dry Creek Road at
MP 2.5-3.5.
Figwort Family.

528
Cordylanthus capitatus I
Clustered Bird's Beak (10-20 in.)
Open places at higher elevations east
of the crest of the Cascade Mountains.
MID JULY: top of Monte Carlo.
Figwort Family.

529*
Digitalis purpurea   III
Foxglove   (2-6 ft.)
Roadsides and forest openings at low
to middle elevations in the west Gorge.
EARLY JUNE: I-84 (westbound) at
Moffett Creek.
LATE JUNE: SR-14 at Cape Horn.
MID JULY: Larch Mountain Corridor.
Figwort Family.

530
Euphrasia stricta   I
Eyebright   (2-4 in.)
MID JULY: thus far found in the Gorge
only on the open flats at the top of
Angels Rest.
Figwort Family.

531
Gratiola ebracteata   II
Bractless Hedge-hyssop   (2-8 in.)
Vernal ponds, wet meadows, and
transitory streams at low to middle
elevations between Bingen and Horse-
thief Lake State Park.
EARLY-MID JUNE: ditch beside Marsh
Cut-off Road.
MID-LATE JULY: Tom McCall Nature
Preserve; meadows on the Major
Creek Plateau.
Figwort Family.

532
Gratiola neglecta IV
Common American (3-10 in.)
Hedge-hyssop
Wet places, especially Columbia River
shores and bottomlands, as far east as
The Dalles. Blooming season depends
on river level.
MID AUG: Mirror Lake.
MID SEPT: Columbia River shores at
Rooster Rock State Park and at Dalton
Point.
Figwort Family.

533
Limosella aquatica II
Mudwort (0-2 in.)
Mudflats along the Columbia River
throughout the Gorge.
MID SEPT: Dalton Point; Rooster Rock
State Park.
Figwort Family.

534*
Linaria dalmatica III
Dalmatian Toad-flax (1-3 ft.)
Roadsides and railroad embankments
at low elevations in the west and
middle Gorge.
EARLY JUNE: SR-14 from the White
Salmon River to Lyle.
Figwort Family.

535*
Linaria vulgaris II
Common Toad-flax, (6-24 in.)
Butter-and-eggs
Roadsides, railroads, and other dis-
turbed areas as far east as Wind Moun-
tain, mostly at low elevations.
MID JULY: Larch Mountain Highway;
Cape Horn; Home Valley Park.
Figwort Family.

536
Lindernia anagallidea I
Slender False Pimpernel (1-6 in.)
Shores of the Columbia, Sandy, and
Deschutes rivers.
EARLY-MID AUG: west bank of the
Deschutes River.
Figwort Family.

537
Lindernia dubia III
Common False Pimpernel (1-6 in.)
Columbia River shores and bottom-
lands throughout the Gorge.
MID AUG: Mirror Lake.
MID SEPT: Dalton Point.
Figwort Family.

538*
Mazus japonicus   II
Mazus   (2-5 in.)
Columbia River bottomlands west of
Bonneville Dam.
LATE AUG-EARLY SEPT: Mirror Lake;
Dalton Point; mouth of Bridal Veil
Creek.
Figwort Family.

539
Mimulus alsinoides   V
Chickweed Monkey Flower   (1-8 in.)
Moist banks and cliffs as far east as
Wishram.
LATE APRIL: Oneonta Gorge; Latourell
Falls; Dog Creek Falls.
Figwort Family.

540
Mimulus breviflorus   I
Short-flowered   (1-5 in.)
Monkey Flower
Drying streambeds, seeps, and ponds
in the east Gorge.
MID-LATE MAY: Tom McCall Nature
Preserve; near Catherine Creek.
Figwort Family.

541
Mimulus floribundus II
Purple-stem (4-12 in.)
Monkey Flower
Columbia River shores, especially in
the west Gorge; streams, seeps, and
wet cliffs throughout the Gorge.
EARLY-MID JULY: cliff at Tunnel #5
on SR-14.
LATE SEPT: mouth of Bridal Veil Creek;
beach at Rooster Rock State Park;
Sand Island.
Figwort Family.

542
Mimulus guttatus V
Common Monkey Flower (3-30 in.)
Seeps, springs, ditches, and small
streams at all elevations.
LATE MAY-EARLY JUNE: Lower Tanner
Creek Road; Dog Creek Falls; numer-
ous roadside ditches.
Figwort Family.

543
Mimulus I
jungermannioides
Columbia Monkey Flower (2-6 in.)
Damp cliffs in the east Gorge.
SUMMER: cliffs near MP 95.9 on I-84
(eastbound).
Figwort Family.

544
Mimulus moschatus III
Musk Monkey Flower (3-10 in.)
Marshes and margins of lakes and
streams at all elevations, especially
Columbia River bottomlands.
AUG: Franz Lake.
LATE AUG-EARLY SEPT: Columbia
River shore west of the Corbett Exit
(#22) on I-84; west bank of the
Deschutes River.
Figwort Family.

545
Nothochelone nemorosa III
Turtle Head (1.5-2.5 ft.)
Woods and forest openings in the west
Gorge, especially at middle to high
elevations.
EARLY AUG: Scenic Highway near
Wahkeena Falls; top of Larch Moun-
tain; Indian Springs.
Figwort Family.

546
Orthocarpus attenuatus III
Narrow-leaf Owl-clover (4-12 in.)
Open grassy areas, generally at low
elevations in the middle and east
Gorge.
MAY: Memaloose Rest Area; Tom
McCall Nature Preserve; Bobs Point.
Figwort Family.

547
Orthocarpus hispidus II
Hairy Owl-clover   (4-12 in.)
Damp meadows, vernal ponds, and
ditches at all elevations between Hood
River and The Dalles.
LATE MAY-EARLY JUNE: damp areas
at Hood River Mountain Meadow;
meadows on the Major Creek Plateau.
Figwort Family.

548
Orthocarpus pusillus II
Dwarf Owl-clover   (2-8 in.)
Open slopes and meadows at all
elevations between Dog Mountain
and The Dalles.
EARLY MAY: grassy hillsides along
backroads out of Mosier.
MID MAY: Hood River Mountain
Meadow.
Figwort Family.

549
Pedicularis bracteosa II
var. flavida
Bracted Lousewort   (1-2 ft.)
Subalpine woods and meadows in the
upper reaches of Eagle Creek and
Herman Creek, including the Benson
Plateau.
MID JULY: near Indian Springs; near
Chinidere Mountain; Wahtum Lake.
Figwort Family.

550
Pedicularis contorta   I
Coiled-beak Lousewort   (6-16 in.)
EARLY-MID JULY: open slopes at the top of Silver Star Mountain and nearby summits.
Figwort Family.

551
Pedicularis groenlandica   II
Elephant-head Lousewort   (1-2.5 ft.)
EARLY JULY: marshes and wet meadows at middle elevations in the west and middle Gorge.
Figwort Family.

552
Pedicularis racemosa   III
Sickle-top Lousewort   (8-18 in.)
Woods at middle to high elevations in the west Gorge.
LATE JULY: near Indian Springs.
AUG: Silver Star Mountain Trail.
Figwort Family.

553
Penstemon acuminatus   II
Sand-dune Penstemon   (10-20 in.)
Sandy ground at low elevations east of
The Dalles.
LATE APRIL-EARLY MAY: east of
Avery Gravel Pit.
EARLY-MID MAY: I-84 near the
turnout at MP 92.3; BLM Parcel at
Celilo.
Figwort Family.

554
Penstemon barrettiae   II
Barrett's Penstemon   (8-24 in.)
Rocky areas, talus slopes, and cliffs
between Hood River and Lyle, mostly
at low elevations.
EARLY-MID MAY: cliffs along the Old
Highway; Old US-30 east of Hood
River.
Figwort Family.

555
Penstemon cardwellii   III
Cardwell's Penstemon   (4-12 in.)
Open, often rocky areas at middle to
high elevations as far east as Mt.
Defiance.
LATE JUNE: Larch Mountain Highway;
upper Washougal River Road.
MID JULY: top of Larch Mountain at
Sherrard Point; near Indian Springs.
Figwort Family.

556
Penstemon deustus   I
var. variabilis
Scorched Penstemon   (1-2 ft.)
LATE JUNE-EARLY JULY: open ridge of
the Columbia Hills west of The Dalles
Mountain Road.
Figwort Family.

557
Penstemon fruticosus   II
var. fruticosus
Shrubby Penstemon   (8-16 in.)
Middle and higher elevation slopes of
Mt. Defiance and the summit ridge of
Monte Carlo.
LATE MAY-EARLY JUNE: South Prairie
Road near MP 5.
LATE JUNE-EARLY JULY: Rainy Lake
Road at the south base of Mt. Defi-
ance.
Figwort Family.

558
Penstemon glandulosus   II
Sticky-stem Penstemon   (1-3 ft.)
Open or lightly wooded areas at
higher elevations (above 1,700 ft.)
east of Hood River.
MID-LATE MAY: open woods near the
top of Chenoweth Road.
LATE MAY-EARLY JUNE: Hood River
Mountain Meadow; Tom McCall
Nature Preserve south of US-30.
Figwort Family.

559
Penstemon ovatus II
Broad-leaf Penstemon (1.5-2.5 ft.)
Rocky areas and cliffs at low elevations
as far east as Multnomah Falls.
LATE MAY-EARLY JUNE: west base of
Rooster Rock; SR-14 at Cape Horn
and the Clark-Skamania County Line.
Figwort Family.

560
Penstemon richardsonii V
var. richardsonii
Cut-leaf Penstemon (1-2 ft.)
Dry cliffs and rocky banks throughout
the Gorge, generally at low elevations.
JULY: cliffs just west of Multnomah
Falls; rocks at the old navigation lock
at Bonneville Dam; SR-14 at Tunnel
#1; US-30 at Mayer State Park.
Figwort Family.

561
Penstemon rupicola III
Rock Penstemon (4-8 in.)
Cliffs and rocky outcrops at either low
or high elevations in the west Gorge.
EARLY MAY: SR-14 at Tunnel #5.
MID MAY: Oneonta Gorge; Beacon
Rock.
LATE MAY-EARLY JUNE: Hamilton
Mountain Trail.
LATE JUNE: top of Larch Mountain.
Figwort Family.

562
Penstemon serrulatus　　　V
Cascade Penstemon　　　(8-24 in.)
Moist places in open woods at all
elevations in the west Gorge.
LATE JUNE: SR-14 at Cape Horn;
Scenic Highway at Bridal Veil Falls
State Park, Latourell Falls, Wahkeena
Falls, and Horsetail Falls.
MID JULY: top of Larch Mountain.
Figwort Family.

563
Penstemon subserratus　　　IV
Fine-tooth Penstemon　　　(1-2.5 ft.)
Rocky places at middle to high eleva-
tions in the west and middle Gorge,
but also low elevations between
Stevenson and the White Salmon River
on the Washington side.
LATE MAY: SR-14 at MP 46.6 and at
the south face of Wind Mountain;
Binns Hill Road.
MID JUNE: Hamilton Mountain Trail;
Grassy Knoll Trailhead.
Figwort Family.

564
Rhinanthus crista-galli　　　II
Yellow Rattle　　　(6-24 in.)
Pastures, fencelines, ditches, and other
moist open places in the west end of
the Gorge; thus far found only on the
Washington side.
LATE MAY-EARLY JUNE: Gibson Road
at the Turner Road junction (MP 3.3);
roads leading to Mt. Zion, especially
Strunk Road.
Figwort Family.

565
Scrophularia lanceolata   I
Lance-leaf Figwort   (2-5 ft.)
Ditches, vernal ponds, and other moist
places from Dog Mountain to the east
end of the Gorge.
MID JUNE: SR-14 near Horsethief
Butte; spring along Hwy 206, 5.7 miles
east of Celilo.
Figwort Family.

566
Synthyris reniformis   II
Snow Queen   (2-6 in.)
Coniferous woods at low elevations,
especially on the Washington side as
far east as Dog Mountain.
LATE MAR-EARLY APRIL: woods
along SR-14 at Cape Horn and near
the Wind River Highway; Eagle Creek
Forest Camp.
Figwort Family.

567
Synthyris stellata   III
Columbia Kittentails   (4-12 in.)
Generally north-facing shaded banks,
cliffs, and ridges in the west Gorge,
chiefly on the Oregon side.
LATE MAR-EARLY APRIL: Scenic
Highway from the Mt. Hood National
Forest boundary to Horsetail Falls.
Figwort Family.

568
Tonella tenella IV
Small-flowered Tonella (4-8 in.)
Open woods between Multnomah
Falls and The Dalles, mostly on the
Oregon side.
MID APRIL: Stanley Rock; Viento
Rocks; Major Creek Road.
Figwort Family.

569*
Verbascum blattaria III
Moth Mullein (2-4 ft.)
Roadsides, fields, and other disturbed
areas throughout the Gorge. Flowers
can be yellow or pink.
JULY: SR-14 east of Washougal near
MP 20 and at the Clark-Skamania
County Line; I-84 at Rooster Rock
State Park.
Figwort Family.

570*
Verbascum thapsus IV
Common Mullein (2-6 ft.)
Roadsides, fields, and other disturbed
areas as far east as The Dalles area.
JULY: SR-14 at the Clark-Skamania
County Line and near Bingen; I-84 at
Rooster Rock State Park.
Figwort Family.

571
Veronica americana V
American Speedwell (6-20 in.)
Columbia River shores and bottom-
lands, roadside ditches, and wet
meadows at all elevations.
MAY-SEPT: Mirror Lake; Grant Lake.
AUG-SEPT: Columbia River shore from
Dalton Point to the mouth of Bridal
Veil Creek.
Figwort Family.

572*
Veronica anagallis- III
aquatica
Water Speedwell (10-40 in.)
Low elevation ditches and slow
streams, especially Columbia River
shores, throughout the Gorge.
LATE MAY: stream along the lower
section of The Dalles Mountain Road.
EARLY SEPT: Columbia River shore
between Dalton Point and the mouth
of Bridal Veil Creek.
Figwort Family.

573
Veronica peregrina III
var. xalapensis
Purslane Speedwell (4-10 in.)
Wet meadows, vernal ponds, swales,
and Columbia River shores throughout
the Gorge.
MID-LATE MAY: Tom McCall Nature
Preserve.
LATE AUG-EARLY SEPT: Columbia
River shore between Dalton Point and
the mouth of Bridal Veil Creek.
Figwort Family.

574
Veronica scutellata II
Marsh Speedwell (6-12 in.)
Wet meadows, vernal ponds, and
Columbia River bottomlands.
LATE JUNE: roadside ditch along
Bristol Road at Laws Corner.
AUG: Beacon Rock Pond; Tom McCall
Nature Preserve; Stevens Pond.
LATE AUG-EARLY SEPT: Mirror Lake.
Figwort Family.

BROOMRAPE FAMILY

575
Orobanche fasciculata II
Clustered Broomrape (3-6 in.)
Dry open ground in the east Gorge.
LATE MAY-EARLY JUNE: flats near
Horsethief Butte.
Broomrape Family.

576
Orobanche ludoviciana II
Sand-dune Broomrape (2-6 in.)
Sandy areas at low elevations in the
east Gorge.
EARLY-MID JUNE: dunes at The Dalles
Airport; Avery Gravel Pit; Crates Point
Dunes.
Broomrape Family.

577
Orobanche pinorum   I
Pine Broomrape   (3-12 in.)
Woods and brushy areas in the middle
and east Gorge, usually associated
with the shrub, Ocean Spray (see
picture #244).
MID JULY: sometimes seen on the
lower sections of Wyeth Trail and
Wygant Trail.
Broomrape Family.

578
Orobanche uniflora   III
var. purpurea
Naked Broomrape   (3-6 in.)
Open or lightly wooded grassy slopes
at low to middle elevations, especially
between Hood River and The Dalles
area.
MID APRIL: Tom McCall Nature
Preserve; SR-14 at MP 70.4; Mema-
loose Viewpoint.
The smaller *var. minuta* occupies
about the same range.
Broomrape Family.

PLANTAIN FAMILY

579
Plantago patagonica   II
Indian-wheat   (2-8 in.)
Dry sandy areas at low elevations in
the east Gorge.
LATE MAY-EARLY JUNE: dunes at The
Dalles Airport; Horsethief Butte; The
Dalles Bridge Road (US-197).
Plantain Family.

580
Galium aparine   V
var. echinospermum
Annual Bedstraw, Cleavers   (4-24 in.)
Open woods and thickets at low
elevations.
EARLY APRIL: SR-14 east of Lyle
Tunnel (MP 77-81).
MAY: Latourell Falls.
Madder Family.

581
Galium boreale   II
Northern Bedstraw   (8-16 in.)
Open woods as far east as Horsetail
Falls.
EARLY JULY: near Crown Point; top of
Angels Rest; Scenic Highway just east
of Horsetail Falls; Gorge Trail east of
Horsetail Falls.
Madder Family.

582
Galium oreganum   III
Oregon Bedstraw   (6-16 in.)
Moist woods at middle to high eleva-
tions in the west Gorge.
MID JUNE: Larch Mountain Highway
east of MP 3.4; Brower Road.
LATE JULY: Spring Camp Road.
Madder Family.

583
Galium triflorum   III
Fragrant Bedstraw   (2-8 in.)
Woods at low to middle elevations in
the west and middle Gorge.
JUNE: Multnomah Falls; Buck Creek
Road.
LATE JUNE: Major Creek Plateau.
JULY: Larch Mountain Corridor.
Madder Family.

HONEYSUCKLE FAMILY

584
Linnaea borealis   IV
Twin Flower   (4-7 in.)
Woods at all elevations in the west
Gorge.
JUNE: Buck Creek Road.
JULY: Larch Mountain Highway.
LATE JULY: Spring Camp Road.
Honeysuckle Family.

585
Lonicera ciliosa   II
Orange Honeysuckle   (vine)
Open woods and thickets in the west
and middle Gorge.
LATE MAY: SR-14 near MP 19.5;
Hamilton Mountain Trail.
EARLY JUNE: Scenic Highway at
Ainsworth State Park.
Honeysuckle Family.

586
Lonicera involucrata   II
Black Twinberry   (3-9 ft.)
Moist woods at middle to high eleva-
tions in the west Gorge, usually at the
margins of lakes and marshes.
MID-LATE JUNE: Rainy Lake.
Honeysuckle Family.

587
Lonicera utahensis   I
Red Twinberry   (3-6 ft.)
MID-LATE JUNE: thus far found in the
Gorge only near the top of Big Huckle-
berry Mountain.
Honeysuckle Family.

588
Sambucus cerulea   III
Blue Elderberry   (6-12 ft.)
Woods and open areas throughout the
Gorge, generally at low elevations.
LATE JUNE: Ainsworth State Park;
Mayer State Park; SR-14 between
Tunnel #5 and the White Salmon
River.
Honeysuckle Family.

589
Sambucus racemosa III
var. arborescens
Red Elderberry (6-15 ft.)
Damp open woods in the west Gorge.
MID APRIL: I-84 from the Sandy River
to Rooster Rock State Park; Scenic
Highway at Ainsworth State Park.
LATE JUNE: top of Larch Mountain.
Honeysuckle Family.

590
Symphoricarpos albus V
Common Snowberry (2-6 ft.)
Open woods at low to middle eleva-
tions in the west and middle Gorge.
JUNE: SR-14 at MP 19.5 and at Grant
Lake; Scenic Highway at Latourell Falls
and Bridal Veil.
Honeysuckle Family.

591
Symphoricarpos mollis II
Creeping Snowberry (6-16 in.)
Open woods at middle to high eleva-
tions between Beacon Rock and the
White Salmon River, most common on
the Washington side.
LATE MAY-EARLY JUNE: Buck Creek
Road.
Honeysuckle Family.

592
Viburnum edule   II
Moosewood Viburnum   (3-9 ft.)
Marshes and streambanks at middle to
high elevations in the west Gorge.
MID-LATE JUNE: Rainy Lake.
Honeysuckle Family.

593
Viburnum ellipticum   III
Oval-leaf Viburnum   (3-9 ft.)
Open woods at low elevations in the
west Gorge.
LATE MAY: Scenic Highway at Bridal
Veil; SR-14 at MP 19.5 and at Grant
Lake.
Honeysuckle Family.

VALERIAN FAMILY

594
Plectritis ciliosa   II
Long-spurred Plectritis   (4-12 in.)
Open or lightly wooded slopes at low
elevations between Hood River and
The Dalles.
EARLY APRIL: slopes above SR-14 east
of Lyle Tunnel (MP 77-81).
MID APRIL: Tom McCall Nature
Preserve; MP .95 on Rock Creek Road.
Valerian Family.

595
Plectritis congesta V
Rosy Plectritis (4-18 in.)
Vernally moist open slopes and mead-
ows in the west and middle Gorge.
LATE APRIL: Starvation Creek State
Park; Viento State Park south of I-84;
Signal Rock.
EARLY MAY: grassy or mossy openings
at Oneonta Bridge and other points
along the Scenic Highway.
Valerian Family.

596
Plectritis macrocera V
White Plectritis (2-18 in.)
Vernally moist open slopes in the east
Gorge.
EARLY APRIL: Stanley Rock; Mema-
loose Rest Area; Horsethief Butte.
EARLY MAY: Tom McCall Nature
Preserve.
LATE MAY: Hood River Mountain
Meadow.
Valerian Family.

597
Valeriana scouleri IV
Scouler's Heliotrope (6-20 in.)
Moist open woods, cliffs, and mossy
banks in the west Gorge, mostly on the
Oregon side.
EARLY MAY: Scenic Highway from the
Mt. Hood National Forest boundary to
Horsetail Falls; Lower Tanner Creek
Road; McCord Creek Falls Trail.
Valerian Family.

598
Valeriana sitchensis   III
Mountain Heliotrope   (1-2 ft.)
Moist places at high elevations in the west Gorge, especially the high ridges at the heads of Eagle Creek and Herman Creek.
LATE JUNE-EARLY JULY: primitive road (gated) above Wahtum Lake.
MID-LATE JULY: near Indian Springs.
Valerian Family.

TEASEL FAMILY

599*
Dipsacus sylvestris   II
Teasel   (3-6 ft.)
Roadside ditches and Columbia River shores in the west and middle Gorge.
EARLY JULY: I-84 near The Dalles.
EARLY AUG: I-84 east of the Sandy River; Lower Tanner Creek Road.
Teasel Family.

CUCUMBER FAMILY

600
Marah oreganus   V
Big Root   (vine)
Open grassy slopes, fields, and rocky areas throughout the Gorge, generally at low elevations.
MID APRIL: SR-14 east of Lyle Tunnel (MP 77-81).
EARLY MAY: Dog Creek Falls.
MID MAY: Cape Horn.
EARLY JUNE: Corbett Hill Road; Crown Point.
Cucumber Family.

601
Campanula rotundifolia   IV
Round-leaf Bluebell   (6-24 in.)
Open or lightly wooded slopes and
cliffs in the west Gorge, typically at low
elevations.
LATE JULY: cliffs near Crown Point;
Cape Horn; Beacon Rock.
Bluebell Family.

602
Campanula scouleri   III
Scouler's Bluebell   (4-10 in.)
Open woods in the west and middle
Gorge, most common at middle to
high elevations.
JULY: Larch Mountain Highway near
the Brower Road junction; South
Prairie Road at the Big Lava Bed.
MID AUG: Spring Camp Road; top of
Larch Mountain.
Bluebell Family.

603
Downingia elegans   III
Showy Downingia   (2-16 in.)
Vernal ponds, pond margins, and
ditches between Mosier and Horsethief
Butte.
EARLY JUNE: Old Highway near the
Lyle-Appleton Road junction.
MID JULY: ponds at the Tom McCall
Nature Preserve; Stevens Pond.
Bluebell Family.

604
Downingia yina I
Willamette Downingia (1-5 in.)
EARLY JUNE: vernal ponds in Horse-
thief Lake State Park and adjacent
lands.
Bluebell Family.

605
Githopsis specularioides II
Common Blue Cup (1-5 in.)
Dry, open or lightly wooded slopes in
the east Gorge, mainly between Hood
River and Lyle.
EARLY MAY: slopes above the Old
Highway at Catherine Creek; grassy
slopes above Major Creek and at
Campbell Creek.
Bluebell Family.

606
Heterocodon rariflorum II
Heterocodon (2-6 in.)
Drying edges of vernal ponds, streams,
and meadows, mostly between the
White Salmon River and The Dalles.
LATE JUNE-EARLY JULY: vernal ponds
at the Tom McCall Nature Preserve.
Bluebell Family.

SECTION FIVE

607
Agoseris aurantiaca   I
Orange Agoseris   (4-14 in.)
Open ridges at high elevations in the
west Gorge.
EARLY-MID JULY: Big Huckleberry
Mountain; Chinidere Mountain.
Composite Family-Chicory Tribe.

608
Agoseris grandiflora   III
Large-flowered Agoseris   (6-18 in.)
Open places east of Bonneville Dam.
Flowers close by mid-morning on
sunny days.
LATE MAY-EARLY JUNE: Tom McCall
Nature Preserve; The Dalles Mountain
Road; Catherine Creek.
MID JUNE: Dry Creek Road at MP 3.5.
Composite Family-Chicory Tribe.

609
Agoseris heterophylla   III
Annual Agoseris   (1-8 in.)
Open areas at all elevations east of the
Wind River. Flowers usually close by
early afternoon.
EARLY APRIL: Spearfish Lake Park;
Horsethief Butte.
MID MAY: top of The Dalles Mountain
Road.
Composite Family-Chicory Tribe.

610*
Cichorium intybus   IV
Chicory   (1-4 ft.)
Roadsides and other disturbed areas as far east as Wishram.
MID JULY: Old US-30 east of Hood River; SR-14 from Home Valley to Bingen; I-84 at Rooster Rock State Park.
Composite Family-Chicory Tribe.

611
Crepis barbigera   II
Bearded Hawksbeard   (1-2 ft.)
Dry grassy places at all elevations east of Mosier.
EARLY-MID JUNE: Tom McCall Nature Preserve.
LATE JUNE: Columbia Hills west of The Dalles Mountain Road.
Composite Family-Chicory Tribe.

612*
Crepis capillaris   III
Smooth Hawksbeard   (1-3 ft.)
Roadsides, pastures, and other disturbed areas in the west and middle Gorge.
EARLY JULY: Snowden Road out of White Salmon.
LATE JULY: roadsides on the Major Creek Plateau; SR-14 at the Clark-Skamania County Line.
Composite Family-Chicory Tribe.

613
Crepis intermedia   II
Gray Hawksbeard   (1-2 ft.)
Open places and lightly wooded areas
from the Little White Salmon River to
Wishram.
MID MAY: Old Highway just west of
the gravel pit near MP 2.4.
LATE MAY: Memaloose Viewpoint;
top of The Dalles Mountain Road.
Composite Family-Chicory Tribe.

614
Crepis occidentalis   II
Western Hawksbeard   (6-16 in.)
Dry open ground from The Dalles area
to the east end of the Gorge.
EARLY JUNE: top of The Dalles Moun-
tain Road; Haystack Butte Road.
Composite Family-Chicory Tribe.

615
Hieracium albiflorum   III
White-flowered Hawkweed   (1-3 ft.)
Open woods at all elevations in the
west and middle Gorge.
MID JULY: Lower Tanner Creek Road.
LATE JULY-EARLY AUG: Spring Camp
Road.
MID AUG: top of Larch Mountain.
Composite Family-Chicory Tribe.

616
Hieracium longiberbe　　Ⅲ
Long-beard Hawkweed　　(1-2 ft.)
Dry cliffs and rocky banks in the west
Gorge, generally at low elevations.
MID JUNE: rocky points on the Colum-
bia River shore at Wyeth; Pillars of
Hercules.
LATE JUNE-EARLY JULY: Scenic High-
way at Shepperd's Dell and Multno-
mah Falls.
MID JULY: top of Angels Rest.
Composite Family-Chicory Tribe.

617
Hieracium scouleri　　Ⅱ
var. scouleri
Scouler's Hawkweed　　(8-36 in.)
Dry open places at all elevations in the
west and middle Gorge.
MID JUNE: Bear Creek Road.
LATE JUNE-EARLY JULY: Nestor Peak;
Monte Carlo.
LATE JULY-EARLY AUG: top of Larch
Mountain.
The *var. cynoglossoides* is found in the
east Gorge.
Composite Family-Chicory Tribe.

618
Hieracium umbellatum　　Ⅰ
Narrow-leaf Hawkweed　　(2-4 ft.)
MID AUG: thus far found in the Gorge
only near the trailhead at Beacon Rock.
Composite Family-Chicory Tribe.

619*
Lapsana communis IV
Nipplewort (1-3 ft.)
Open woodlands in the west Gorge,
generally at low elevations.
MID JULY: Scenic Highway at Latourell
Falls and Shepperd's Dell; Lower
Tanner Creek Road.
Composite Family-Chicory Tribe.

620
Microseris borealis I
Bog Microseris (8-16 in.)
MID JULY: marshes at middle to high
elevations west of Bonneville Dam.
Blooms in mornings.
Composite Family-Chicory Tribe.

621
Microseris laciniata I
Cut-leaf Microseris (1-2 ft.)
Moist open slopes west of Bonneville
Dam. Best seen in the morning when
flowers are open.
EARLY JUNE: near the top of Hamilton
Mountain Trail; Aldrich Butte.
Composite Family-Chicory Tribe.

622
Microseris lindleyi   II
Lindley's Annual   (6-16 in.)
Microseris
Open places at low to middle eleva-
tions between Bingen and Horsethief
Butte, commonly on the Washington
side. Best seen in the morning when
flowers are open.
MID APRIL: SR-14 east of Lyle Tunnel
(MP 77-81).
LATE APRIL: grassy slopes along the
Old Highway at Catherine Creek.
Composite Family-Chicory Tribe.

623
Microseris nutans   III
Nodding Microseris   (8-18 in.)
Open woods, often near oaks, in the
middle Gorge. Best seen in the morning
when flowers are open.
MID MAY: high on Vensel, Carroll,
and Chenoweth roads between Mosier
and The Dalles.
EARLY JUNE: Grassy Knoll Trailhead;
DNR Forest; Hood River Mountain
Meadow.
Composite Family-Chicory Tribe.

624
Microseris troximoides   III
False Agoseris   (4-20 in.)
Dry open places at all elevations east
of the Little White Salmon River. Best
seen in the morning.
EARLY APRIL: Old Highway at Cather-
ine Creek.
MID APRIL: Stanley Rock; Memaloose
Viewpoint.
MID MAY: top of Carroll Road and
Vensel Road; Haystack Butte.
Composite Family-Chicory Tribe.

625
Prenanthes alata   II
Western Rattlesnake Root   (6-18 in.)
Moist shady places in the west Gorge;
commonly on high ridges on the
Washington side.
EARLY AUG: trail near Upper Horsetail
Falls.
MID AUG: south trail to Silver Star
Mountain.
Composite Family-Chicory Tribe.

626*
Sonchus oleraceus   III
Common Sow-thistle   (1-3 ft.)
Blooms throughout the summer in
roadsides and other disturbed areas,
mostly in the west Gorge.
MID JULY: SR-14 at the Clark-Skama-
nia County Line; Lower Tanner Creek
Road.
Composite Family-Chicory Tribe.

627
Stephanomeria paniculata   II
Stiff-branch Skeletonweed   (1-3 ft.)
Dry areas at low elevations in the east
Gorge, often in roadsides.
MID AUG: SR-14 east of Bingen.
LATE AUG-EARLY SEPT: rocky areas
near The Dalles Bridge on the Oregon
side.
Composite Family-Chicory Tribe.

628
**Stephanomeria tenuifolia II
var. tenuifolia**
Narrow-leaf Skeletonweed (8-24 in.)
Cliffs and talus slopes at low elevations
between Wind Mountain and Dalles-
port, mainly on the Washington side.
AUG: SR-14 at the base of Dog Moun-
tain and at Tunnel #1.
Composite Family-Chicory Tribe.

629*
Tragopogon dubius III
Yellow Salsify (1-2.5 ft.)
Roadsides and dry open slopes
throughout the Gorge.
EARLY JUNE: I-84 near the Sandy
River and near Starvation Creek State
Park; I-84 (westbound) near Moffett
Creek; SR-14 between Wind Mountain
and Bingen.
Composite Family-Chicory Tribe.

630*
Arctium minus II
Common Burdock (2-4 ft.)
Roadsides and disturbed areas at low
elevations as far east as Cascade Locks.
EARLY AUG: Multnomah Falls;
Ainsworth State Park; Beacon Rock.
Composite Family-Thistle Tribe.

631*
Centaurea cyanus   V
Bachelor's Button   (1-3 ft.)
Roadsides and fields at low elevations
in the west and middle Gorge.
MID JUNE: SR-14 between Dog
Mountain and Bingen; US-30 at
Mayer State Park.
Composite Family-Thistle Tribe.

632*
Centaurea diffusa   V
Tumble Knapweed   (1.5-2.5 ft.)
Roadsides, railroads, and other dry
disturbed ground.
EARLY JULY: SR-14 east of Horsethief
Butte and near MP 89.7.
LATE JULY: I-84 at Wyeth, Viento
State Park, and Stanley Rock; US-30 at
Tom McCall Nature Preserve.
Composite Family-Thistle Tribe.

633*
Centaurea maculosa   III
Spotted Knapweed   (2-4 ft.)
Roadsides and other disturbed areas,
generally at low elevations in the west
and middle Gorge.
MID-LATE JULY: I-84 east of Starva-
tion Creek State Park; Old US-30 east
of Hood River.
Composite Family-Thistle Tribe.

634*
Centaurea solstitialis II
Yellow Star Thistle (1-2.5 ft.)
Roadsides and disturbed areas at low
elevations east of Bingen.
AUG: SR-14 east of Bingen; Bobs
Point.
Composite Family-Thistle Tribe.

635*
Cirsium arvense IV
Canada Thistle (2-5 ft.)
Fields and roadsides in the west and
middle Gorge.
EARLY JULY: Grant Lake.
MID JULY: Scenic Highway at Crown
Point, Multnomah Falls, and Ainsworth
State Park.
Composite Family-Thistle Tribe.

636
Cirsium callilepis I
(Cirsium centaureae)
Mountain Thistle (2-3 ft.)
JULY-AUG: thus far found in the
Gorge only along the Pacific Crest Trail
near Chinidere Mountain.
Composite Family-Thistle Tribe.

637
Cirsium remotifolium II
Weak Thistle (2-4 ft.)
Meadows and open hillsides at middle
to high elevations in the middle Gorge.
LATE JUNE-EARLY JULY: upper section
of the Dog Mountain Trail.
Composite Family-Thistle Tribe.

638
Cirsium undulatum III
Wavy-leaf Thistle (2-5 ft.)
Dry open areas from Dallesport to the
east end of the Gorge.
LATE JUNE: Horsethief Butte.
MID JULY: top of The Dalles Mountain
Road.
Composite Family-Thistle Tribe.

639*
Cirsium vulgare IV
Bull Thistle (2-4 ft.)
Fields, roadsides, and other disturbed
areas in the west and middle Gorge.
LATE JULY: I-84 east of the Sandy
River; Scenic Highway at Ainsworth
State Park; SR-14 east of Washougal
and at the Clark-Skamania County
Line.
Composite Family-Thistle Tribe.

640
Brickellia grandiflora   II
Thoroughwort   (1.5-2.5 ft.)
Dry open woods at scattered locations
in the west and middle Gorge.
LATE AUG: Wyeth Trailhead; Horsetail
Falls Trail.
Composite Family-Thoroughwort
Tribe.

641
Anaphalis margaritacea   V
Pearly Everlasting   (1-3 ft.)
Roadsides and forest openings at all
elevations in the west and middle
Gorge.
EARLY AUG: Lower Tanner Creek
Road; Beacon Rock.
LATE AUG: top of Larch Mountain.
Composite Family-Everlasting Tribe.

642
Antennaria dimorpha   III
Low Pussytoes   (1-3 in.)
Dry open places in the east Gorge.
EARLY APRIL: Horsethief Butte.
MID APRIL: Tom McCall Nature
Preserve.
Composite Family-Everlasting Tribe.

643
Antennaria luzuloides   II
Woodrush Pussytoes   (8-20 in.)
Open woods at middle to high eleva-
tions on both sides of the Columbia
River between Hood River and The
Dalles.
EARLY JUNE: various locations along
Hood River Mountain Road, including
Hood River Mountain Meadow.
Composite Family-Everlasting Tribe.

644
Antennaria microphylla   II
Rosy Pussytoes   (6-16 in.)
Dry rocky places in open woods and
meadows at middle to high elevations
in the middle Gorge.
MID JUNE: Grassy Knoll Trailhead.
MID JULY: Indian Mountain.
Composite Family-Everlasting Tribe.

645
Antennaria neglecta   II
var. attenuata
Field Pussytoes   (6-16 in.)
Forest openings in the west Gorge,
usually at low elevations.
EARLY MAY: Signal Rock.
LATE MAY: top of Angels Rest; Hamil-
ton Mountain Trail; near Crown Point.
Closely-related *var. howellii* is also
found in the west Gorge.
Composite Family-Everlasting Tribe.

646
Antennaria racemosa III
Slender Pussytoes (4-18 in.)
Moist forest openings at middle to
high elevations in the west Gorge.
LATE MAY-EARLY JUNE: Grassy Knoll
Trailhead.
LATE JUNE-EARLY JULY: Indian Moun-
tain; Warren Lake Trail.
Composite Family-Everlasting Tribe.

647
Gnaphalium chilense III
Cotton-batting Plant (1-2.5 ft.)
Moist ground at low elevations
throughout the Gorge, often in ditches
and other disturbed areas.
AUG: Mirror Lake; Horsethief Lake
State Park.
SEPT: Dalton Point.
Composite Family-Everlasting Tribe.

648
Gnaphalium II
microcephalum
Slender Cudweed (1-2.5 ft.)
Dry open places, often on roadsides, at
low elevations in the middle and east
Gorge.
LATE AUG-EARLY SEPT: SR-14 near
Dog Mountain, near Major Creek (MP
72.6), and near Horsethief Butte.
Composite Family-Everlasting Tribe.

649
Gnaphalium palustre   III
Lowland Cudweed   (2-6 in.)
Damp ground throughout the Gorge,
including vernal ponds and shores of
the Columbia River.
JUNE: Tom McCall Nature Preserve.
AUG: Dalton Point; Columbia River
shore west of the Corbett Exit (#22).
Composite Family-Everlasting Tribe.

650
Gnaphalium purpureum   I
Purplish Cudweed   (8-16 in.)
Columbia River bottomlands in the
west Gorge.
LATE JULY-EARLY AUG: mouth of
Lindsey Creek.
Composite Family-Everlasting Tribe.

651
Psilocarphus elatior   I
Tall Woolly Heads   (2-4 in.)
Vernal ponds and drying mud of roads
and ditches at low to middle elevations
between Dog Mountain and The
Dalles.
JUNE: Hood River Mountain Meadow.
Composite Family-Everlasting Tribe.

652
Aster alpigenus |
Alpine Aster   (4-8 in.)
LATE JUNE-EARLY JULY: wet meadows
at higher elevations (over 3,000 ft.) in
the west Gorge.
Composite Family-Aster Tribe.

653
Aster ascendens |
(Aster chilensis subspecies
adscendens)
Long-leaf Aster   (10-20 in.)
SEPT: thus far found in the Gorge only
on the banks of the Deschutes River.
Composite Family-Aster Tribe.

654
Aster eatonii ||
Eaton's Aster   (2-4 ft.)
Damp open ground in the east Gorge.
LATE AUG: ditch along SR-14 at
Horsethief Butte; west bank of the
Deschutes River; banks of the Klickitat
River; I-84 (eastbound) at MP 95.2.
Composite Family-Aster Tribe.

655
Aster foliaceus var. parryi   I
Leafy Aster   (1-2 ft.)
AUG: thus far found in the Gorge only
in open areas at the top of Dog Moun-
tain and nearby Cook Hill.
Composite Family-Aster Tribe.

656
Aster frondosus   I
Alkali Aster   (1-12 in.)
Margins of streams and ponds at low
elevations east of The Dalles.
EARLY SEPT: west bank of the Des-
chutes River.
Composite Family-Aster Tribe.

657
Aster glaucescens   II
Klickitat Aster   (2-5 ft.)
Open woods, mostly at middle to high
elevations between the Wind River
and the White Salmon River. Thus far
found only on the Washington side.
EARLY AUG: Nestor Peak; Monte
Carlo; Big Huckleberry Mountain;
South Prairie Road between MP 5-10.
Composite Family-Aster Tribe.

658
Aster ledophyllus II
Cascade aster (1-2 ft.)
Open woods at the heads of Eagle
Creek and Herman Creek, and at Silver
Star Mountain.
MID AUG: Indian Springs Road.
Composite Family-Aster Tribe.

659
Aster modestus II
Great Northern Aster (1-3 ft.)
Damp ground along streams, lakes,
and marshes in the west and middle
Gorge.
LATE JULY: Wind River shore.
EARLY SEPT: Rainy Lake; Wahtum
Lake.
Composite Family-Aster Tribe.

660
Aster occidentalis II
var. intermedius
Western Aster (1-2 ft.)
Damp meadows and margins of vernal
ponds in the east Gorge.
LATE JULY-EARLY AUG: Stevens
Pond; Tom McCall Nature Preserve.
LATE AUG-EARLY SEPT: Columbia
River shore at the Deschutes River
mouth.
Composite Family-Aster Tribe.

661
Aster occidentalis
var. occidentalis
Western Aster   (6-14 in.)
Damp meadows at high elevations in
the west Gorge.
SEPT: thus far found in the Gorge only
at a damp meadow along the trail
between Rainy Lake and North Lake.
Composite Family-Aster Tribe.

662
Aster oregonensis
Oregon White-top Aster   (2-4 ft.)
Dry open woods at low elevations in
the west Gorge.
LATE AUG-EARLY SEPT: Gillette Lake.
Composite Family-Aster Tribe.

663
Aster pansus
Tufted White Prairie Aster   (1-3 ft.)
Margins of drying vernal ponds east of
The Dalles.
LATE AUG-EARLY SEPT: thus far
found in the Gorge only at a vernal
pond just west of Horsethief Lake
State Park.
Composite Family-Aster Tribe.

664
Aster radulinus   II
Rough-leaf Aster   (6-24 in.)
Dry open woods between Crown
Point and Wishram.
MID AUG: Eagle Creek Overlook
Picnic Area; Aldrich Butte; top of
Angels Rest.
Composite Family-Aster Tribe.

665
Aster subspicatus   V
Douglas' Aster   (1-4 ft.)
Moist meadows, ditches, and stream
margins, especially shores and bottom-
lands of the Columbia River.
LATE AUG-EARLY SEPT: Dalton Point;
Mirror Lake; Scenic Highway at
Ainsworth State Park; Beacon Rock
Shore.
Composite Family-Aster Tribe.

666
Chrysopsis villosa   IV
Hairy Goldaster   (6-20 in.)
Sandy ground at low elevations east of
Dog Mountain; particularly abundant
in dunes.
EARLY JULY: SR-14 at Tunnel #1 and
east of Lyle Tunnel (MP 77-81); dunes
near The Dalles Airport; near The
Dalles Bridge Ponds.
Composite Family-Aster Tribe.

667
Chrysothamnus IV
nauseosus var. albicaulis
Gray Rabbit Brush (1.5-6 ft.)
Dry plains at low elevations, mainly
east of The Dalles.
LATE SEPT-EARLY OCT: The Dalles
Bridge Road (US-197); Hwy 206
between Celilo and the Deschutes
River.
Composite Family-Aster Tribe.

668
Chrysothamnus II
viscidiflorus
var. lanceolatus
Green Rabbit Brush (1.5-4 ft.)
Dry areas at low elevations east of The
Dalles area.
SEPT: Horsethief Butte; Bobs Point.
Composite Family-Aster Tribe.

669
Erigeron annuus IV
Annual Fleabane (2-5 ft.)
Moist ground in roadsides and waste
places at low elevations in the west
Gorge.
MID JULY: Lower Tanner Creek Road;
SR-14 at the Clark-Skamania County
Line; Cape Horn Road; Grant Lake.
SEPT: Scenic Highway near MP 23-24.
Composite Family-Aster Tribe.

670
Erigeron divergens   II
Spreading Fleabane   (6-18 in.)
Dry places and waste areas at low elevations in the east Gorge.
EARLY JUNE: Locke Lake; Bobs Point.
Composite Family-Aster Tribe.

671
Erigeron filifolius   III
Thread-leaf Fleabane   (6-20 in.)
Dry areas at all elevations east of Crates Point.
EARLY JUNE: Crates Point Dunes; Horsethief Butte; Hwy 206 near Celilo.
Composite Family-Aster Tribe.

672
Erigeron howellii   II
Howell's Daisy   (8-20 in.)
Rocky slopes facing north or east at all elevations in the west Gorge, mostly on the Oregon side.
LATE MAY: Multnomah Falls.
EARLY JULY: top of Larch Mountain.
Composite Family-Aster Tribe.

673
Erigeron linearis   I
Yellow Desert Daisy   (2-12 in.)
Open areas at the top of the Columbia
Hills.
EARLY JUNE: Columbia Hills ridge
west of The Dalles Mountain Road;
east side of Haystack Butte.
Composite Family-Aster Tribe.

674
Erigeron oreganus   II
Columbia Gorge Daisy   (2-6 in.)
Under overhanging basalt cliffs in the
west Gorge, generally on the Oregon
side.
MID JUNE: Oneonta Gorge; Latourell
Falls; McCord Creek Falls.
Composite Family-Aster Tribe.

675
Erigeron peregrinus   I
var. eucallianthemus
Subalpine Daisy   (1-2 ft.)
EARLY-MID JULY: open slopes at the
top of Silver Star Mountain and nearby
summits.
Composite Family-Aster Tribe.

676
Erigeron philadelphicus   I
Pink Fleabane   (1-3 ft.)
Wet areas along streams at low elevations in the east Gorge.
LATE MAY: Old Highway near a prominent rock outcrop about 1.5 miles east of Major Creek.
JUNE: west bank of the Deschutes River.
Composite Family-Aster Tribe.

677
Erigeron poliospermus   II
var. poliospermus
Cushion Fleabane   (3-6 in.)
Dry areas at low elevations between Dallesport and Horsethief Butte.
MID APRIL: Horsethief Butte; Spearfish Lake Park.
Composite Family-Aster Tribe.

678
Erigeron subtrinervis   I
var. conspicuus
Three-vein Fleabane   (1-2 ft.)
MID-JULY: open areas high on Dog Mountain.
Composite Family-Aster Tribe.

679
Grindelia columbiana III
Columbia River Gumweed   (1-2.5 ft.)
Ditches, lakes, vernal ponds, and
Columbia River shores in the east
Gorge.
MID JULY: Spearfish Lake Park; Bobs
Point; The Dalles Bridge Ponds; lower
section of The Dalles Mountain Road;
junction of SR-14 with The Dalles
Bridge Road (US-197).
Composite Family-Aster Tribe.

680
Grindelia nana II
var. integrifolia
Low Gumweed   (1-2.5 ft.)
Dry streambeds at low elevations east
of The Dalles.
LATE AUG: mouth of the Deschutes
River; mouth of Fulton Canyon.
Composite Family-Aster Tribe.

681
Gutierrezia sarothrae I
Match Brush   (8-18 in.)
Dry slopes and rocky areas at low
elevations in the east end of the Gorge.
EARLY-MID SEPT: south shore of
Miller Island; near the road at Bobs
Point.
Composite Family-Aster Tribe.

682
Haplopappus   II
carthamoides
Rayless Goldenweed   (1-2 ft.)
Dry ground between Hood River and
Wishram, generally at middle to high
elevations.
MID-LATE JULY: Hood River Mountain
Meadow; top of The Dalles Mountain
Road; Mayer State Park.
Composite Family-Aster Tribe.

683
Haplopappus hallii   III
Hall's Goldenweed   (1-2 ft.)
Dry open areas at all elevations in the
middle and east Gorge.
LATE SEPT: Dog Creek Falls; Tom
McCall Nature Preserve; Hood River
Mountain Meadow; Old US-30 west
of Mosier.
Composite Family-Aster Tribe.

684
Haplopappus resinosus   III
Columbia Goldenweed   (1-2 ft.)
Basalt cliffs overlooking the Columbia
River in the east Gorge.
EARLY SEPT: Old Highway; SR-14 east
of Lyle; Bobs Point.
Composite Family-Aster Tribe.

685
Machaeranthera   II
canescens
Hoary Aster   (1-2.5 ft.)
Dry open places at low elevations from
The Dalles area to the east end of the
Gorge.
SEPT: dunes near The Dalles Airport;
Hwy 206 east from Celilo.
Composite Family-Aster Tribe.

686
Solidago canadensis   IV
Meadow Goldenrod   (2-5 ft.)
Moist soil of meadows, open woods,
and roadsides throughout the Gorge.
LATE AUG-EARLY SEPT: Dalton Point;
Lower Tanner Creek Road; Home
Valley Park; I-84 east of The Dalles.
Composite Family-Aster Tribe.

687
Solidago gigantea   I
Late Goldenrod   (2-5 ft.)
Damp areas in the east Gorge.
MID-LATE SEPT: west bank of the
Deschutes River; occasionally at
Beacon Rock shore and Pierce Island.
Composite Family-Aster Tribe.

688
Solidago occidentalis III
Western Goldenrod (3-6 ft.)
Damp ground along shores and bot-
tomlands of the Columbia and Des-
chutes rivers.
MID SEPT: Dalton Point; Grant Lake;
west bank of the Deschutes River;
Beacon Rock shore.
Composite Family-Aster Tribe.

689
Solidago spathulata I
var. neomexicana
Little Goldenrod (6-14 in.)
Gravelly open ground.
MID-LATE JULY: thus far found in the
Gorge only at the top of Angels Rest.
Composite Family-Aster Tribe.

690
Adenocaulon bicolor V
Pathfinder (1-3 ft.)
Coniferous woods in the west Gorge.
MID JULY: Beacon Rock; Latourell
Falls; McCord Creek Falls Trail.
Composite Family-Senecio Tribe.

691
Arnica amplexicaulis   II
var. amplexicaulis
Streambank Arnica   (1-2.5 ft.)
Streambanks and wet meadows,
especially at middle to high elevations
in the west Gorge.
JUNE: Wind River shore.
Composite Family-Senecio Tribe.

692
Arnica amplexicaulis   III
var. piperi
Columbia Gorge Arnica   (2-3 ft.)
Waterfalls and wet cliffs in the west
Gorge.
MID JUNE: Dog Creek Falls.
MID JULY: Horsetail Falls; Latourell
Falls.
Composite Family-Senecio Tribe.

693
Arnica cordifolia   II
Heart-leaf Arnica   (8-18 in.)
Open woods between the Little White
Salmon River and The Dalles.
MID-LATE MAY: Hood River Mountain
Road; Carroll Road and Vensel Road;
DNR Forest.
Composite Family-Senecio Tribe.

694
Arnica discoidea II
Rayless Arnica (1-2 ft.)
Open woods at all elevations in the
middle Gorge.
MID JUNE: Buck Creek Road.
LATE JUNE: South Prairie Road.
MID JULY: near Indian Springs; Rainy
Lake Road near Mt. Defiance.
Composite Family-Senecio Tribe.

695
Arnica latifolia IV
Broad-leaf Arnica (1-2 ft.)
Coniferous woods and forest openings
at all elevations in the west Gorge.
LATE MAY: Beacon Rock.
MID JULY: Indian Springs; top of Larch
Mountain; Rainy Lake Road.
Composite Family-Senecio Tribe.

696
Crocidium multicaule V
Gold Stars (2-7 in.)
Dry open places at all elevations
throughout the Gorge.
LATE MAR: Spearfish Lake Park; Dog
Creek Falls; Stanley Rock; Memaloose
Viewpoint.
LATE APRIL-EARLY MAY: Oneonta
Bridge.
Composite Family-Senecio Tribe.

697
Luina nardosmia   II
Silver Crown Luina   (1.5-3.5 ft.)
Open woods, often near oaks, usually
at middle to high elevations from
Viento State Park to The Dalles.
MID MAY: high on the Carroll, Vensel,
and Chenoweth roads between Mosier
and The Dalles.
MID JUNE: top of Monte Carlo.
Composite Family-Senecio Tribe.

698
Petasites frigidus   III
Sweet Colt's Foot   (1-2 ft.)
Streams, ditches, and waterfalls at low
elevations in the west Gorge.
LATE MAR-EARLY APRIL: Scenic
Highway east of Crown Point; Lower
Tanner Creek Road.
Composite Family-Senecio Tribe.

699
Senecio bolanderi   IV
var. harfordii
Columbia Gorge   (6-24 in.)
Groundsel
Moist coniferous woods at all eleva-
tions in the west Gorge.
MID JUNE: Scenic Highway at Shep-
perd's Dell, Latourell Falls, and Multno-
mah Falls.
Composite Family-Senecio Tribe.

700
Senecio cymbalarioides I
Alpine Meadow Groundsel (1-2 ft.)
JULY: wet meadows at higher eleva-
tions in the west Gorge.
Composite Family-Senecio Tribe.

701
Senecio foetidus II
Sweet Marsh Groundsel (1-3.5 ft.)
Wet meadows at higher elevations in
the east Gorge.
MID JUNE: meadows on the Major
Creek Plateau; Laws Corner.
Composite Family-Senecio Tribe.

702
Senecio hydrophilus II
Great Swamp Groundsel (1.5-3 ft.)
Rocky and gravelly shores of the
Columbia River as far east as Bonneville
Dam.
LATE JULY-EARLY AUG: Dalton Point;
mouth of Wahkeena Creek.
Composite Family-Senecio Tribe.

703
Senecio integerrimus   III
var. exaltatus
Yellow Western Groundsel   (1-2.5 ft.)
Open woods and forest openings as
far east as The Dalles, usually at middle
to high elevations.
LATE APRIL-EARLY MAY: Mayer State
Park.
MID MAY: high on the Carroll, Vensel,
and Dry Creek roads.
Composite Family-Senecio Tribe.

704
Senecio integerrimus   III
var. ochroleucus
White Western Groundsel   (1-2.5 ft.)
Open woods from the Wind River to
the Klickitat River, often at low eleva-
tions.
LATE APRIL-EARLY MAY: SR-14 at
Dog Mountain and at the tunnels (MP
58-60); Major Creek Road.
LATE MAY-EARLY JUNE: Hood River
Mountain Meadow.
Composite Family-Senecio Tribe.

705*
Senecio jacobaea   III
Tansy Ragwort   (2-4 ft.)
Roadsides, railroads, and pastures at
low elevations in the west Gorge.
EARLY AUG: pastures and railroad
embankments along I-84 east of the
Sandy River; Ainsworth State Park.
LATE AUG: Lower Tanner Creek Road.
Composite Family-Senecio Tribe.

706
Senecio triangularis   III
Arrow-leaf Groundsel   (2-4 ft.)
Streambanks and marshes in the west
Gorge, generally at middle to high
elevations.
LATE JUNE-EARLY JULY: Wind River
shore.
LATE JULY-EARLY AUG: North Lake
Trailhead; Indian Springs Road .5 mile
south of Wahtum Lake Forest Camp.
Composite Family-Senecio Tribe.

COMPOSITE FAMILY-ANTHEMIS TRIBE

707
Achillea millefolium   V
Yarrow   (1-2 ft.)
Open places, roadsides, and waste
ground at all elevations throughout
the Gorge.
LATE MAY-EARLY JUNE: Horsethief
Butte.
MID JUNE: Dog Creek Falls.
LATE JUNE-EARLY JULY: Scenic High-
way at Crown Point and Latourell Falls.
Composite Family-Anthemis Tribe.

708*
Anthemis cotula   III
Mayweed Chamomile   (10-20 in.)
Roadsides, disturbed ground, and
Columbia River shores.
EARLY JUNE: I-84 east of the Sandy
River.
JULY: Ainsworth State Park; Old US-
30 east of Hood River.
Composite Family-Anthemis Tribe.

709
Artemisia biennis   II
Biennial Wormwood   (1-3 ft.)
Columbia River shores throughout the
Gorge.
LATE AUG-EARLY SEPT: mouth of the
Deschutes River.
LATE SEPT: Dalton Point.
Composite Family-Anthemis Tribe.

710
Artemisia campestris   II
var. scouleriana
Northern Wormwood   (1-4 ft.)
Sandy or gravelly places along the
Columbia River, generally above the
high-water mark.
LATE AUG-EARLY SEPT: rocky shore
of the Columbia River near The Dalles
Dam Visitor Center; Hwy 206, two
miles east of Celilo; Bobs Point.
Composite Family-Anthemis Tribe.

711
Artemisia douglasiana   I
Douglas' Wormwood   (2-5 ft.)
Open, usually moist places throughout
the Gorge.
MID JULY: top of Dog Mountain; top
of Cook Hill.
Composite Family-Anthemis Tribe.

712
Artemisia dracunculus II
Dragon Wormwood, (2-4 ft.)
Tarragon
Dry open places at low elevations east
of The Dalles.
EARLY SEPT: Horsethief Butte; Celilo
Park; turnout at MP 92.3 on I-84
(eastbound).
Composite Family-Anthemis Tribe.

713
Artemisia lindleyana III
Riverbank Wormwood (1-2 ft.)
Gravelly and rocky shores of the
Columbia River throughout the Gorge,
generally below the high-water mark.
LATE AUG-EARLY SEPT: Beacon Rock
shore; Bobs Point; mouth of McCord
Creek.
Composite Family-Anthemis Tribe.

714
Artemisia ludoviciana II
Western Wormwood (1-5 ft.)
Dry open areas at low elevations
throughout the Gorge, but commonly
east of Mosier.
LATE JULY-EARLY AUG: Bobs Point;
dunes near The Dalles Airport; The
Dalles Bridge Ponds; west bank of the
Deschutes River.
MID-LATE AUG: pond at MP 94.4 on
I-84 (eastbound).
Composite Family-Anthemis Tribe.

715
Artemisia rigida   I
Rigid Sagebrush   (1-2 ft.)
Dry open places east of The Dalles.
EARLY OCT: east base of Horsethief
Butte; BLM Parcel at Celilo.
Composite Family-Anthemis Tribe.

716
Artemisia suksdorfii   II
Suksdorf's Wormwood   (2-4 ft.)
Open areas at low elevations west of
Cascade Locks.
MID-LATE JUNE: Scenic Highway east
of Multnomah Falls; Lower Tanner
Creek Road; Woodward Creek Bridge
near the Beacon Rock shore.
Composite Family-Anthemis Tribe.

717
Artemisia tilesii   III
Mountain Wormwood   (3-8 ft.)
Roadsides and other low elevation
areas in the west Gorge as well as high
mountain ridges in the west Gorge.
LATE SEPT-EARLY OCT: I-84 at Dalton
Point and Rooster Rock State Park.
Composite Family-Anthemis Tribe.

718
Artemisia tridentata   II
Big Sagebrush   (4-8 ft.)
Dry areas at low elevations, chiefly
east of The Dalles.
EARLY OCT: I-84 at the Mayer State
Park overpass (MP 75.3); west bank of
the Deschutes River.
Composite Family-Anthemis Tribe.

719*
Chrysanthemum   V
leucanthemum
Ox-eye Daisy   (1-2.5 ft.)
Roadsides, fields, and disturbed areas
west of the Klickitat River.
EARLY JUNE: I-84 at Cascade Locks
and Starvation Creek State Park; Grant
Lake.
MID JULY: top of Larch Mountain.
Composite Family-Anthemis Tribe.

720
Hymenopappus filifolius   I
Columbia Cut Leaf   (1-2 ft.)
Dry sandy areas at low elevations in
the east end of the Gorge.
EARLY JUNE: near the south shore of
Miller Island.
Composite Family-Anthemis Tribe.

721

Matricaria matricarioides   II
Pineapple Weed   (2-10 in.)
Roadsides and waste places through-
out the Gorge.
LATE MAY: Beacon Rock shore; top of
The Dalles Mountain Road.
Composite Family-Anthemis Tribe.

722*

Tanacetum vulgare   III
Common Tansy   (2-5 ft.)
Roadsides, fields, and other disturbed
areas.
EARLY AUG: I-84 at Rooster Rock
State Park and Dalton Point; Scenic
Highway between Multnomah Falls
and Ainsworth State Park.
Composite Family-Anthemis Tribe.

COMPOSITE FAMILY-SUNFLOWER TRIBE

723

Balsamorhiza deltoidea   V
Northwest Balsamroot   (1-3 ft.)
Open areas and oak woods, especially
east of Bingen.
MID APRIL: Old Highway; Lyle-
Appleton Road.
EARLY MAY: Hood River Mountain
Meadow.
LATE MAY: top of Aldrich Butte.
Composite Family-Sunflower Tribe.

724
Balsamorhiza sagittata   II
Arrow-leaf Balsamroot   (1-2.5 ft.)
Generally at middle to high elevations
in the Columbia Hills.
MID MAY: top of The Dalles Mountain
Road.
Composite Family-Sunflower Tribe.

725
Bidens cernua   III
Nodding Beggar Ticks   (6-36 in.)
Columbia River shores and bottom-
lands throughout the Gorge. Also
found at upland ponds in the east
Gorge.
AUG: Tom McCall Nature Preserve.
EARLY-MID SEPT: Mirror Lake; Dalton
Point.
Composite Family-Sunflower Tribe.

726
Bidens frondosa   III
Leafy Beggar Ticks   (6-36 in.)
Columbia River shores and bottomlands throughout the Gorge. Also in
damp ground at upland ponds in the
east Gorge.
SEPT: mouth of Bridal Veil Creek; SR-
14 at Rock Creek Lake; Beacon Rock
shore.
Composite Family-Sunflower Tribe.

727
Chaenactis douglasii   II
var. douglasii
Hoary False Yarrow   (6-18 in.)
Dry open places in the east Gorge.
LATE MAY: The Dalles Bridge Road
(US-197); Horsethief Butte.
MID JUNE: top of The Dalles Mountain
Road.
Composite Family-Sunflower Tribe.

728
Coreopsis atkinsoniana   IV
Columbia Coreopsis   (1-4 ft.)
Columbia River shores throughout the
Gorge.
JULY: Beacon Rock shore; Grant Lake;
shoreline east of The Dalles Riverside
Park.
Composite Family-Sunflower Tribe.

729
Eriophyllum lanatum   III
var. integrifolium
Woolly Sunflower,   (6-12 in.)
Oregon Sunshine
Dry open, often rocky places in the
east Gorge.
MID-LATE MAY: US-30 at Mayer
State Park; I-84 from Rowena to The
Dalles; SR-14 east of Lyle Tunnel (MP
77-81).
Composite Family-Sunflower Tribe.

730
Eriophyllum lanatum V
var. lanatum
Woolly Sunflower, (10-24 in.)
Oregon Sunshine
Dry open areas and forest openings, mostly in the west Gorge.
EARLY-MID JUNE: Scenic Highway from Crown Point to Multnomah Falls; I-84 at Cascade Locks; SR-14 at Beacon Rock and at MP 46.6.
Composite Family-Sunflower Tribe.

731
Gaillardia aristata III
Blanket Flower (1-2 ft.)
Open grassy places from Dog Mountain to the east end of the Gorge, primarily at low elevations on the Washington side.
MID JUNE: SR-14 west of Murdock; The Dalles Bridge Road (US-197); The Dalles Mountain Road; Horsethief Lake State Park.
Composite Family-Sunflower Tribe.

732
Helenium autumnale IV
var. grandiflorum
Sneezeweed (1-3 ft.)
Columbia River shores and bottomlands throughout the Gorge.
LATE AUG-EARLY SEPT: Mirror Lake; Dalton Point; Beacon Rock shore.
Composite Family-Sunflower Tribe.

733
Helianthella uniflora   I
Little Sunflower   (2-4 ft.)
Open woods and brushy areas at
middle to high elevations east of Dog
Mountain; thus far found only on the
Washington side.
EARLY-MID JULY: top of Monte Carlo.
Composite Family-Sunflower Tribe.

734
Helianthus annuus   II
Common Sunflower   (2-6 ft.)
Open places at low elevations, mainly
east of The Dalles.
LATE JULY-EARLY AUG: The Dalles
Mountain Road; SR-14 east of
Wishram.
LATE AUG-EARLY SEPT: Hwy 206
between Celilo and the Deschutes
River.
Composite Family-Sunflower Tribe.

735
Lagophylla ramosissima   III
Rabbit Leaf   (4-20 in.)
Dry open slopes at all elevations
between the White Salmon River and
Horsethief Butte.
EARLY JUNE: Catherine Creek; The
Dalles Bridge Road (US-197).
Composite Family-Sunflower Tribe.

736
Lasthenia glaberrima I
Smooth Lasthenia (2-10 in.)
Vernal ponds between Mosier and
Horsethief Lake State Park.
LATE MAY: Stevens Pond.
EARLY JUNE: Horsethief Lake State
Park and adjacent lands.
Composite Family-Sunflower Tribe.

737
Layia glandulosa II
White Tidy Tips (6-12 in.)
Open sandy ground in the east Gorge.
LATE APRIL: Miller Island and hills on
the east side of the Deschutes River.
Composite Family-Sunflower Tribe.

738
Madia citriodora II
Lemon-scented Tarweed (6-16 in.)
Dry hillsides in the east Gorge.
MID-LATE MAY: Tom McCall Nature
Preserve.
EARLY JUNE: top of The Dalles Moun-
tain Road.
Composite Family-Sunflower Tribe.

739
Madia exigua III
Little Tarweed (2-10 in.)
Dry open woods and grasslands from
Viento State Park to Wishram.
EARLY JUNE: Mayer State Park; Old
Highway at Major Creek; Hood River
Mountain Meadow.
Composite Family-Sunflower Tribe.

740
Madia glomerata II
Stinking Tarweed (1-2 ft.)
Dry open, often disturbed areas,
mainly in the middle Gorge.
MID JULY: north side of the Little
Wind River Bridge.
LATE JULY-EARLY AUG: ponds east of
Mosier; SR-14 at Rock Creek Lake;
meadows on the Major Creek Plateau.
Composite Family-Sunflower Tribe.

741
Madia gracilis IV
Common Tarweed (2-24 in.)
Dry open places throughout the
Gorge, but chiefly between Wind
Mountain and The Dalles.
MID MAY: Catherine Creek; Mema-
loose Viewpoint; Tom McCall Nature
Preserve. The closely related, but
larger and coarser species, *Madia
sativa*, grows along roadsides in the
west and middle Gorge. Blooms in
August at Ainsworth State Park.
Composite Family-Sunflower Tribe.

742
Rigiopappus leptocladus ll
Bristle Head (4-12 in.)
Dry grassy slopes at all elevations in
the east Gorge.
MID MAY: Old Highway at Catherine
Creek; Tom McCall Nature Preserve.
Composite Family-Sunflower Tribe.

743
Wyethia amplexicaulis ll
Northern Wyethia (1-2.5 ft.)
Vernally moist, open or lightly wooded
areas in the east Gorge, usually at
middle to high elevations.
LATE MAY: Old Highway near the
Lyle-Appleton Road junction.
MID JUNE: Hood River Mountain
Road; meadows on the Major Creek
Plateau.
Closely related *Wyethia angustifolia* is
also found along the Old Highway.
Composite Family-Sunflower Tribe.

744
Xanthium strumarium lll
Common Cocklebur (6-36 in.)
Columbia River shores and bottom-
lands throughout the Gorge, and also
vernal ponds in the east Gorge.
EARLY SEPT: Beacon Rock shore;
Rooster Rock State Park shore; Dalton
Point; The Dalles Bridge (Oregon side).
Composite Family-Sunflower Tribe.

RECOMMENDED TRIPS: A SAMPLER

THE TRIPS suggested below will lead the wildflower explorer into some of the interesting corners of the Gorge—into forests, by streams and waterfalls, through pine-oak woodlands, across open grasslands, and along river shores. For the beginner, these trips will serve as an introduction to the vast natural garden of the Columbia Gorge. For the experienced amateur botanist, they will help target times to view favorite plants throughout the Gorge.

All the places are easily accessed, and only two of the trips, Hamilton Mountain Trail and Upper Multnomah Basin, require considerable hiking.

A great many places are not included here for several reasons, primarily for lack of space. Other reasons are: difficulty of access of one sort or another; private land ownership considerations, especially when sites are near residences or farms; and sensitivity of an area to visitor impact. Some places are so sensitive, in fact, that the author, far from inviting others to visit, refrains from visiting these areas himself so as to minimize the unnecessary disturbance.

The selected dates for these trips, arbitrarily arranged on a weekly schedule, are based on records maintained by the author and a colleague, Lois Kemp, for over ten years. They indicate probable "prime" blooming times for at least some of the species named. While there will be yearly variations in blooming times because of weather conditions, the reader should be able to visit an area a week before or after the suggested date and still see most of the plants in bloom. Remember also that the listed plants are only examples of what to expect on a given date. Normally, other colorful species will also be flowering at the same time.

The number listed in parentheses following each flower name is a picture number and refers the reader to a detailed description contained in the Guide section. In most cases, MP refers to mileage markers on the side of the road, but in some instances where a road has no mileage markers, MP refers to actual miles traveled.

MARCH 10

Some Gorge flowers must be seen early or not at all. For these plants, April may be too late.

EAST OF LYLE TUNNEL (SR-14 at MP 77-81)   On these south-facing slopes, Columbia Desert Parsley (388) and Salt-and-pepper (397) are in bloom, along with Gold Stars (696), Grass Widows (47), Yellow Bells (28), and Smooth Prairie Stars (207). If it's a warm day, check skin and clothes for ticks before leaving the field. A warm day in early spring is prime time for ticks in the east Gorge, and this particular area is a likely spot. Old-timers say that eating garlic will help repel the pests.

TOP OF THE DALLES MOUNTAIN ROAD   The Dalles Mountain Buttercup (147), a bright yellow flower, and white Canby's Desert Parsley (387) are found here at the top of the Columbia Hills ridge, together with tiny flowers such as Spring Whitlow-grass (180) and Scale Pod (183).

MARCH 24-OCTOBER 1

SCENIC HIGHWAY   This highway, particularly the stretch between Corbett and Dodson, provides a great flower show throughout the spring and well into summer, enough for a whole set of trips. Parking is scarce in places, and walking along the road can be risky where there are no shoulders, so take care. In late March, Western Wake-robin (37) is common in the woods, and blue Columbia Kittentails (567) adorn the mossy banks along the road between the Mt. Hood National Forest boundary (MP 19) and Ainsworth State Park (MP 23).

By early April, Slender Toothwort (177), wild Bleeding Heart (162), and Oregon Wood Sorrel (327) bloom along the roadside, especially at Ainsworth State Park. As the season progresses, tall Poison Larkspur (137) and Western Solomon Plume (31) appear in the woods.

On the open slopes next to the Scenic Highway, May brings Rosy Plectritis (595), Rough Wallflower (181), clumps of Field Chickweed (111), and Red Columbine (127).

By June, these same slopes are bright with Oregon Sunshine (730), Broad-leaf Stonecrop (199), and Cascade Penstemon (562).

Finally, in mid-July, Round-leaf Bluebells (601) drape from the cliffs near Crown Point and Multnomah Falls, along with Oregon Stonecrop (198). After July, the rush to bloom subsides and continues at a more subdued pace. By November, only blue-tinged Annual Fleabane (669) remains in bloom along the Scenic Highway at Ainsworth State Park.

APRIL 1

In the east Gorge, things are going strong by April 1.

MAJOR CREEK ROAD   It is best to park the car along the paved Old Highway and walk the unpaved Major Creek Road; there is very little traffic on this 1.75-mile dead-end road. After a two hundred-foot climb the road levels out, winding through pine-oak woodlands with occasional rock outcrops and meadows in a lovely quiet valley. Just about every wildflower species of the Major Creek area can be found somewhere along this road. At least eighty-five native wildflowers bloom here in the spring, including Glacier Lilies (24), Sierra Snake Root (405), Cut-leaf Violet (349), Upland Larkspur (135), and Western Buttercup (145).

KLICKITAT FISH LADDER   A few miles east of Major Creek lies the Klickitat River Valley. The area near the fish ladder is reserved for fishing by members of the Yakima Indian Nation, but non-Indians are permitted to roam the area in search of wildflowers. In the pine-oak woods, Great Hound's Tongue (487) is abundant, along with Western Buttercup (145) and Oaks Toothwort (176). On the rocks near the Klickitat River (take care!), the rare Violet Suksdorfia (223) can be found, along with Smooth Prairie Star (207), Gray-leaf Desert Parsley (393), and Pungent Desert Parsley (390), named for its strong characteristic aroma. Here, near the river, one can also find the two other species of Prairie Star that grow in the Gorge.

BOBS POINT   Around April 1, many of the species blooming at the Klickitat Fish Ladder, such as Desert Shooting Star (429), are also blooming here. But many other species can be found as well, among them Golden Currant (228), Long-leaf Phlox (466), Smooth Desert Parsley (391), and some species that grow in sandy ground, such as Veiny Dock (92) and Pale Wallflower (182).

APRIL 8

MCCORD CREEK FALLS   From the parking area at Yeon State Park, the trail to the top of the falls is about a mile long and climbs to around five hundred feet in elevation. In early April, Red-flowering Currant (234) blooms near the trailhead. Along the trail, which gradually climbs through the woods, keep an eye out for Deer's-head Orchid (50). Higher up, Columbia Kittentails (567) and Western Saxifrage (221) grow on the steep mossy banks. On the cliffs near the top of the falls, Smooth-leaf Douglasia (436) and Cliff Paintbrush (522) will be starting to bloom, the first in a series of beautiful wildflowers that find a home on these cliffs.

LOWER TANNER CREEK ROAD   The earliest flowers blooming along this short road include Indian-plum (246), Salmonberry (269), Stream Violet (343), Sweet Colt's Foot (698), Bleeding Heart (162), and Western Wake-robin (37).

Western Corydalis (160) will be just getting started, to peak in early May. Sitka Mist Maidens (479) may also be just starting to blossom on the cliffs next to the road. It, too, will not peak until early May. Like McCord Creek, Lower Tanner Creek cannot be properly appreciated unless it is visited several times during the spring and early summer.

APRIL 15

STARVATION CREEK STATE PARK   Along the short trail to the base of Starvation Creek Falls, there should still be some Dutchman's Breeches (161) blooming, although it will be past its prime. To the west of the parking lot, on the trail to Mt. Defiance, Cascade Rock Cress (163) will be in bloom on the steep slopes, generally under the shelter of trees or brush.

TOM MCCALL NATURE PRESERVE   Here and in adjacent Mayer State Park, a great variety of flowers is in bloom in mid-April, although some, like Columbia Desert Parsley (388), are just about finished flowering. Mertens' Saxifrage (220) can be seen on the low cliffs beside the road. On the open grassy plateau, look for Upland Larkspur (135), Poet's Shooting Star (434), Naked Broomrape (578), Rigid Fiddleneck (482), Northwest Balsamroot (723), Gray-leaf Desert Parsley (393), and Bare-stem Desert Parsley (396). Among the smaller flowers are two kinds of Plectritis, White Plectritis (596) and Long-spurred Plectritis (594), and tiny White Meconella (158).

SR-14—US-197 JUNCTION   Where The Dalles Bridge Road meets SR-14, Northwest Balsamroot (723), Broad-leaf Lupine (299), and Hood River Milk-vetch (279) can be seen blooming in mid-April.

HORSETHIEF BUTTE   On the flats at the base of Horse-thief Butte, Bitter Brush (257) should be coming into bloom, together with the almost shrub-like Puccoon (490), flat clusters of Cushion Fleabane (677), Panicled Death-camas (42), and Thread-leaf Phacelia (476). The tall white spikes of Thick-leaf Thelypody (192) can be seen on the cliffs above.

HAMILTON MOUNTAIN TRAIL   For the first mile or two, the trail passes through typical Douglas-fir forest, containing plants such as Baneberry (122), Cascade Oregon Grape (154), Candy Flower (105), Nevada Pea (287), Fairy Bells (22), and blue-flowered Oregon Anemone (126). Farther up, steep, grassy hillsides (called "hanging meadows") host Gold Stars (696), Rosy Plectritis (595), and Harsh Paintbrush (520). On cliffs near the trail, there are clumps of Smooth-leaf Douglasia (436) and Spreading Phlox (464). A month later, the show will be totally different, but just as colorful.

HOOD RIVER MOUNTAIN MEADOW   This huge meadow needs most of the day for close inspection. At this point in the season, there should still be some Lance-leaf Spring Beauties (95) left, as well as some Poet's Shooting Stars (434), and maybe some Glacier Lilies (24), but Ball-head Waterleaf (470) will be abundant in the shade of the oaks, and Northwest Balsamroot (723) will be near its prime. Fern-leaf Desert Parsley (389) and Nine-leaf Desert Parsley (399) will also be in bloom, along with Big-head Clover (317) and False Agoseris (624).

ONEONTA BRIDGE   Here, at MP 22 on the Scenic Highway, Cliff Larkspur (134) blooms on the mossy walls, along with Scouler's Heliotrope (597), Rosy Plectritis (595), Sitka Mist Maidens (479), and masses of the little yellow Chickweed Monkey Flower (539).

CATHERINE CREEK   Bright purple masses of Barrett's Penstemon (554) are seen on cliffs beside the Old Highway, just west of Catherine Creek. In rocky areas at Catherine Creek, Bitter Root flowers (97) lie almost flat on the ground. Here and there are clumps of Douglas' Buckwheat

(76) and the white-flowered vines of Big Root (600). Also scattered over the area are Baby Stars (458) and blue-and-white Bicolored Cluster Lilies (15), although sometimes the blue is very pale. Sand Clover (319), Small-head Clover (318), and White-tip Clover (320) are in bloom here; also Spurred Lupine (300) and Meadow Death-camas (43).

MAY 12

THE DALLES MOUNTAIN ROAD   Near the bottom of this eight-mile road, which leads to the top of the Columbia Hills, look for Hood River Milk-vetch (279). Several different kinds of lupine can be seen along the road, including Broad-leaf Lupine (299), Whitish Lupine (303), and the little Prairie Lupine (301). About halfway up, there are Yakima Milk-vetch (281) and Choke Cherry (256), lots of Bare-stem Desert Parsley (396) and Slender-fruited Desert Parsley (392), and Arrow-leaf Balsamroot (724). At the top of the ridge are Douglas' Buckwheat (76), Showy Phlox (467), and Hood's Phlox (465), the latter beginning to fade.

Before heading home, stop at Horsethief Butte to see Hoary False Yarrow (727), Pearhip Rose (263), Western Blue Flag (44), and Thread-leaf Fleabane (671).

MAY 19

WASCO BUTTE   From Mosier, take Carroll Road and then Vensel Road south to the crest at 2,200 feet elevation. There one can either turn onto a 0.5-mile spur road which leads north to Wasco Butte, or continue on Vensel Road and turn left down Chenoweth Road toward The Dalles. Among the many flowers along the roads are Nootka Rose (261), Yellow Western Groundsel (703), Few-flowered Pea (288), Red Columbine (127), Showy Phlox (467), Silver Crown Luina (697), Slender Godetia (359), Large-flowered Collomia (453), Suksdorf's Desert Parsley (398), and Sticky-stem Penstemon (558). Columbia Frasera (443) can be seen on the lower section of Chenoweth Road, along with Ball-head Cluster Lily (13). On I-84 between The Dalles and Mosier, note the clumps of Oregon Sunshine (729), particularly near MP 80.

GRASSY KNOLL TRAILHEAD   From Wind River Highway, turn right onto Bear Creek Road (Road 6808) and follow it for eight to ten miles to the junction with Road 68. Turn left on Road 68 and drive approximately two miles to the Grassy Knoll Trailhead, where there are meadows on both sides of the road. Among the wildflowers in bloom are White Western Groundsel (704), Spreading Phlox (464), Sticky Cinquefoil (251), and the ornately fringed Oregon Campion (118). A little farther north on Road 68, Indian-parsnip (382) grows on open rocky slopes above and below the road. In early July, tall Mountain Hollyhock (337) blooms along this road.

FISHER HILL GRADE   From the Klickitat Fish Ladder, continue up the Fisher Hill Grade. Along the way, look for Nevada Deer-vetch (294), Big-head Clover (317), Deer Brush (334), Ball-head Cluster Lily (13), light-blue Columbia Frasera (443), and Hyacinth Cluster Lily (16). At the top of the grade (about 2,000 feet elevation), turn left and return via Lyle-Appleton Road. For an alternate—and longer—return route, turn north on Lyle-Appleton Road and climb to the top of the Major Creek Plateau at Appleton, roughly eleven miles from Lyle. Then drive west and south across the plateau and down to White Salmon. The route passes several wet meadows, most of them privately owned, with many interesting wildflowers.

AIRPORT DUNES   Just north of The Dalles Bridge (on US-197) turn left on The Dalles Airport Road. The dunes are located 0.3 mile up the road on the right (north) side. Find a place to park off the pavement. The dunes are located in the southeast corner of The Dalles Municipal Airport. The plants found here are typical of those found in all dunes areas of the Gorge. Among these are Pale Evening Primrose (376), White Sand-verbena (94), Thread-leaf Fleabane (671),

Hoary Aster (685), Hairy Goldaster (666), Lance-leaf Scurf-pea (310), Wiry Knotweed (88), Sand-dune Broomrape (576), Indian-wheat (579), Prairie Lupine (301), and Oregon Sunshine (729).

HORSETHIEF LAKE STATE PARK   Along the main park road, Blanket Flowers (731) and Wavy-leaf Thistles (638) are blooming.

SR-14 AT MP 91   Prickly-pear Cactus (351) is found on both sides of the road near here.

SR-14 AT MP 91.4 AND MP 93   Look for Showy Milkweed (450).

SR-14 AT MP 94-98   On the slopes above the highway, there are extensive stands of Silky Lupine (308).

SR-14 AT MP 81   The white Velvet Lupine (302) and the Blanket Flower (731) grow along this stretch of highway.

OLD HIGHWAY   Just west of the Lyle-Appleton Road junction, the Old Highway levels out and passes through residential and pasture land. Northern Wyethia (743) and closely related *Wyethia angustifolia* grow along this stretch of highway, but are fading by early June. In wet areas along the north side of the road, Showy Downingia (603) is in bloom.

JUNE 16 ─────────────────────────────

LARCH MOUNTAIN HIGHWAY   Since there are long stretches of this road where there is no place to park safely off the pavement, it is best to find the nearest parking place and walk back. In mid-June, Oregon Flag (46), Cardwell's Penstemon (555), Thicket Deer-vetch (292), Thimbleberry (267), Broad-leaf Lupine (298), and Salal (416) are all seen in bloom along this highway. This "wave" of blooming tends to rise in elevation, moving up the slopes as the season progresses. In mid-June at the top of the Larch Mountain Highway (MP 14.5), look for Round-leaf Violets (346) and Avalanche Lilies (25).

————————————

RAINY LAKE ROAD (ROAD 2820)   From Dee in the Hood River Valley, take the righthand fork at the T-junction. After crossing the West Fork of the Hood River, turn right again at the sign reading "Rainy Lake" and start climbing. At about 3,500 feet elevation, look for Fine-tooth Penstemon (563). At 4,000 feet elevation, along the south base of Mt. Defiance, Shrubby Penstemon (557) can be seen on rocky slopes next to the road, and farther on, Broad-leaf Arnica (695) as well.

RAINY LAKE   The lake lies a short walk from Rainy Lake Road (Road 2820). Typical marsh vegetation is found on the boggy shores at the north and northwest sides of the lake. Here one can find Marsh Violets (347), Marsh-marigolds (128), Tall Mountain Shooting Stars (433), Western False Asphodel (36), Northern Star Flower (439), Moosewood Viburnum (592), Alpine Mitrewort (212), and Black Twinberry (586).

————————————

SOUTH PRAIRIE ROAD (ROAD 66)   Plants of the Heath Family are common at middle to upper elevations in the west Gorge. South Prairie Road offers an easy opportunity to see some of them along the eastern edge of the Big Lava Bed. Between MP 4 and MP 6, several Heath Family members are in bloom at this time, including White-vein Pyrola (425), Candy Stick (410), Salal (416), Pipsissewa (414), and Little Pipsissewa (413). Pinedrops (423), another member of the Heath Family, will bloom here in late July. Other roadside flowers include Rayless Arnica (694), Nevada Deer-vetch (294), Scouler's Bluebell (602), and Twin Flower (584).

MCCORD CREEK FALLS   In the woods along the first part of the trail, look for Indian Pipe (421). Higher up, see Taper-tip Onion (8) and Nodding Onion (10). Cascade Penstemon (562) is in bloom, along with two shrubs, Mock Orange (236) and Ocean Spray (244). Still higher up, Columbia Gorge Daisies (674) hang from the cliffs next to the trail.

MULTNOMAH FALLS   If it's a warm day, venture into the spray from the falls and scan the cliffs for the characteristic leaves and little white flowers of Oregon Sullivantia (224).

JULY 7

UPPER MULTNOMAH BASIN   This area is directly below the viewpoint on Larch Mountain. It has the largest stand of old-growth forest in the Columbia Gorge, perhaps two thousand acres. The easiest approach is via the short Spring Camp Road at MP 11.6 on the Larch Mountain Highway. This road leads to a complex of hiking trails in the Multnomah Basin. Along these trails, Foam Flower (226) is frequently seen on the forest floor, as are Canadian Dogwood (407) and Bead Lily (21). Scattered along the trails are Western Rhododendron (427), False Bugbane (151), Inside-out Flower (156), Columbia Wind Flower (123), Arrow-leaf Groundsel (706), and Mertens' Coral Root (52). Pinesap (418) and Western Twayblade (64) are less commonly seen.

TOP OF LARCH MOUNTAIN   Take the short trail to the viewpoint. Along the trail, look for False Lily-of-the-valley (30), Cascade Mountain Ash (272), Red Elderberry (589), and Starry Solomon Plume (32). At the viewpoint, note Howell's Daisy (672), Martindale's Desert Parsley (395), Bear-grass (41), Cardwell's Penstemon (555), and Cascade Penstemon (562).

JULY 14

WAHTUM LAKE   From Wahtum Lake Forest Camp, follow the trail down to the lake and along the south shore. Among the flowers to be seen are Bracted Lousewort (549), False Bugbane (151), Hall's Isopyrum (138), and Goat's Beard (238).

INDIAN SPRINGS   When wandering about this area or following the trails through the subalpine forest, expect to encounter plants such as Turtle Head (545), Gray's Lovage (386), Sickle-top Lousewort (552), Mountain Heliotrope (598), Mountain Spiraea (275), Tiger Lily (29), and Drummond's Cinquefoil (249). In late August, Explorer's Gen-

283

tian (444) can be seen in bloom along the Pacific Crest Trail in this area and on the nearby Indian Mountain Trail.

AUGUST 4 ——————————————————

YOUNG CREEK AND MIRROR LAKE   In areas that are flooded in spring, but become dry by late summer, look for Yampah (403), which is sometimes up to three feet tall, and Hooded Ladies' Tresses (68), which can reach heights of two feet. In the low areas that remain wet, typical plants are Common Silverweed (248), Field Mint (505), Sneezeweed (732), Water Pepper (86), Common American Hedgehyssop (532), Fringed Loosestrife (437), and Wapato (5).

AUGUST 18 ——————————————————

BEACON ROCK POND   In the wet shoreline areas, one can see Simple-stem Bur Reed (6), Mad-dog Skullcap (512), Northern Bugleweed (504), American Speedwell (571), and Marsh Speedwell (574). In the pond, there are Yellow Water-lilies (120).

BEACON ROCK SHORE   Plants typical of the Columbia River shore are found here, including Columbia Coreopsis (728), Riverbank Wormwood (713), Sneezeweed (732), Common Tansy (722), Meadow Goldenrod (686), Douglas' Aster (665), and Field Mint (505).

AUGUST 25 ——————————————————

COLUMBIA RIVER SHORE FROM DALTON POINT TO BRIDAL VEIL CREEK   This stretch of shore varies from rocky to sandy, to muddy, back to rocky, and so on throughout its 1.6-mile length. In drier areas find Douglas' Aster (665), Western Goldenrod (688), and Licorice Root (284). Great Swamp Groundsel (702) grows on wet, rocky shores. On sandy and muddy shores, typical plants include Western Water-hemlock (381), Common Monkey Flower (542), Common Silverweed (248), Common False Pimpernel (537), Common Willow-herb (366), American Speedwell (571), and Water Speedwell (572).

────────────────────

WEST BANK OF THE DESCHUTES RIVER   The part of the riverbank referred to here is the first mile up from the Highway 206 bridge. Several plants found here are not commonly found in the rest of the Gorge. Among them are Alkali Aster (656), Long-leaf Aster (653), Low Gumweed (680), Late Goldenrod (687), and an introduced species, Wild Tobacco (517). Many other wildflowers found on the banks of the Deschutes, such as Western Goldenrod (688) and Nodding Beggar Ticks (725), are common elsewhere in the Gorge.

OCTOBER 1 ──────────────────────

CELILO   From I-84, take the Celilo Exit (97). Then, take Highway 206 eastward. Along this road, Gray Rabbit Brush (667) and Green Rabbit Brush (668) are in bloom, along with Hoary Aster (685), Common Sunflower (734), Strict Buckwheat (78), and Wiry Knotweed (88).

BLOOMING TIMES

THE NUMBERS listed in this section are picture numbers and refer the reader to the numbered wildflower entries contained in the Guide section. West, Middle, and East refer to specific geographic sections of the Gorge. The boundaries are shown on the fold-out map.

MARCH

WEST			MIDDLE			EAST		
37	132	231	221	696		24	28	47
411	566	567				48	143	147
						166	180	183
						194	207	345
						387	388	391
						397	696	

APRIL

WEST			MIDDLE			EAST		
7	22	26	20	27	135	15	20	27
27	50	66	153	161	167	28	42	72
104	105	108	174	176	181	92	95	101
124	144	150	208	237	255	102	104	135
153	154	161	258	344	370	139	145	153
162	165	166	405	459	487	155	158	166

172	174	177	496	525	568	178	179	182
185	221	234	704			184	186	192
237	242	246				206	207	208
254	269	287				219	220	223
325	327	332				228	230	257
343	348	412				279	280	281
432	436	464				282	283	288
479	501	525				297	299	303
539	566	567				307	317	325
568	589	595				345	349	390
698						392	393	396
						398	429	430
						434	456	457
						459	465	466
						467	470	473
						476	480	482
						487	490	491
						496	498	525
						527	553	554
						578	580	594
						596	600	609
						622	624	642
						677	723	737

MAY

WEST			MIDDLE			EAST		
19	20	27	15	19	24	12	13	15
30	31	32	25	43	45	43	44	53
34	70	73	71	80	95	71	75	76
103	105	111	104	146	157	91	97	139
122	123	126	171	173	181	140	191	204
134	137	149	199	211	233	239	240	256
152	159	160	240	242	251	257	263	279
163	164	168	263	293	295	281	283	288
169	170	175	297	321	323	299	300	303
210	216	217	326	335	370	313	315	316
219	220	225	389	399	404	317	318	319
227	229	240	405	440	476	320	359	375
241	243	247	485	704	721	392	396	398
251	254	256				409	458	466
258	260	261				467	469	474
266	267	270				478	482	483
287	304	305				484	485	486
306	323	335				488	495	497
342	347	378				498	511	526
389	395	399				531	540	546

401	402	404		547	548	553
408	409	416		554	558	572
419	431	435		573	596	605
440	446	454		609	613	623
464	471	472		624	676	693
481	485	489		697	703	721
494	520	522		723	724	727
524	557	559		729	738	741
561	563	564		742	743	
571	580	585				
593	595	597				
600	645	672				
695	696	721				
723						

JUNE ———————————————————

WEST			MIDDLE			EAST		
8	21	23	8	11	14	1	3	8
25	29	30	16	18	32	9	13	14
31	32	33	38	49	51	16	17	43
35	37	41	53	56	59	49	55	74
46	66	69	63	71	80	79	91	97
73	74	83	115	118	120	98	107	117
96	100	105	146	167	190	120	133	140
106	108	109	200	203	233	148	195	197
110	115	116	235	236	237	236	252	294
122	125	127	244	251	252	301	302	308
128	129	136	259	264	291	310	314	331
138	148	149	294	295	300	338	351	359
154	156	162	305	324	329	361	373	376
173	175	181	334	336	337	442	443	449
193	199	201	341	362	401	450	451	453
202	205	212	451	481	534	458	461	463
224	232	237	574	583	584	547	558	565
238	247	252	623	631	644	575	576	579
259	262	266	694	697	704	603	604	606
268	269	272	707			608	611	614
274	277	278				631	638	643
289	292	298				649	651	670
305	311	321				671	673	676
322	324	328				701	707	720
332	336	341				727	731	735
346	348	357				736	738	739
358	382	384				743		
385	395	406						

407	416	420
421	427	428
433	446	454
455	468	485
489	509	510
521	529	542
550	555	557
559	561	562
563	564	571
582	583	585
586	587	588
589	590	591
592	600	616
617	621	623
629	637	646
652	674	691
692	699	708
716	719	730

JULY

WEST			MIDDLE			EAST		
10	21	29	4	39	51	2	4	60
30	31	32	56	57	59	67	77	81
36	41	52	63	80	118	88	94	114
54	58	60	119	286	290	131	142	195
61	62	64	294	296	337	253	276	284
65	66	74	338	340	350	296	301	308
83	96	99	353	374	413	310	330	350
105	106	112	423	447	452	354	355	356
113	116	119	460	462	507	360	373	442
123	127	138	528	551	560	449	499	507
151	156	175	577	610	612	531	543	556
198	202	209	617	632	633	560	599	603
212	213	215	644	678	694	606	610	632
216	217	218	708	711	728	634	638	666
222	224	226	733	740		679	682	728
244	245	249				733		
250	264	265						
268	273	275						
284	285	286						
290	294	296						
298	309	311						
312	328	342						
347	358	363						
364	365	366						
367	377	379						
380	386	400						

402	406	407
410	414	415
416	418	420
421	422	423
424	425	426
438	439	441
447	453	455
462	475	477
485	492	500
506	507	508
509	513	514
521	523	529
530	535	541
549	550	551
552	555	557
560	562	569
570	571	581
582	583	584
598	601	602
607	615	616
619	620	626
635	636	637
639	646	652
659	669	672
675	689	690
691	692	695
700	706	707
708	719	728

AUGUST

WEST

5	6	40
68	69	86
113	114	120
121	130	131
138	142	175
196	198	214
248	271	274
276	296	333
339	366	367
368	369	372
380	381	383
413	414	424
437	444	445
448	493	500
502	504	505
508	512	513

MIDDLE

4	276	296
337	368	372
417	493	502
515	628	648
657	740	

EAST

4	6	78
82	88	90
121	141	195
284	296	368
371	403	442
448	493	502
503	515	516
517	536	543
574	627	647
648	654	660
663	680	681
710	714	725
734		

518 523 532
536 537 544
545 552 571
574 599 601
602 615 617
618 625 630
636 640 641
647 649 650
655 658 659
662 664 665
702 705 706
722 732

SEPTEMBER

WEST

58 68 82
84 85 87
89 90 148
175 187 188
189 248 253
366 380 444
519 532 533
537 538 541
544 571 572
573 574 647
659 661 662
665 669 686
687 688 709
713 725 726
732 744

MIDDLE

4   85 648
686 688 713

EAST

4   85 88
93 141 248
352 371 516
517 544 627
648 653 656
660 663 668
680 681 683
684 685 686
687 688 709
710 712 713
734 744

OCTOBER

WEST

717

MIDDLE

EAST

667 668 715
718

PLACE NAMES

TO SUPPLEMENT the information given in this book, certain maps may be useful, including maps of the Mt. Hood National Forest and the Gifford Pinchot National Forest. "Forest Trails of the Columbia Gorge" is a more detailed map, and a map of the Columbia Gorge National Scenic Area is especially valuable. All these maps are available from the U.S. Forest Service Regional Office in Portland, Oregon.

For serious wildflower exploring in the Gorge, U.S. Geological Survey topographic maps are indispensable. These "topo" maps can be purchased from the U.S. Geological Survey, Denver, Colorado 80225. Write them for the free Index Maps for Washington and Oregon.

Following the heading for each place name entry is a notation giving a location on the fold-out map. On the map, the vertical lines numbered 0 to 19 are spaced at intervals of five minutes (5') of longitude, or about four miles. These lines provide a reference grid for locating places in the Gorge. The prefixes O and W indicate Oregon and Washington sides, respectively. Each entry has its appropriate notation. For example, Larch Mountain has the notation O/3.9, while that of Stacker Butte is W/15.8.

The abbreviation MP is for milepost. In most cases, MP will refer to actual roadside markers, but in cases where there are no markers, MP refers to the distance along a specific road. For a list of other abbreviations, *see* the front inside cover.

AINSWORTH STATE PARK (O/4.3)   Forested area along both sides of Scenic Highway just east of Horsetail Falls (MP 22.5-23.5).

ALDRICH BUTTE (W/5.3)   Former lookout site at 1,000 ft. elev., 1.3 miles east of Hamilton Mountain. Turn north from SR-14 at MP 38.55, then right on Cascade Drive; proceed to end of paved road alongside Greenleaf Slough; walk jeep road (private land) to top. Also accessed from Pacific Crest Trail, although a more direct trail access is being proposed.

ANGELS REST TRAIL (O/2.9)   Popular 2 mile trail to top of Angels Rest at 1,600 ft. elev. Trailhead located on Scenic Highway just west of Bridal Veil (MP 16.5).

ASHES LAKE (W/6.1)   75 acre lake along SR-14 at MP 42.7-43.1; separated from Columbia River by causeway for SR-14 and railroad.

AUGSPURGER MOUNTAIN (W/8.5)   One of the highest points on Washington side of Gorge at 3,670 ft. elev.; head of Dog Creek.

AVERY GRAVEL PIT (W/16.9)   Sandy area near railroad tracks, east of Avery Boat Ramp. Leave SR-14 at MP 89.7; from boat ramp, drive about 1.75 miles east on gravel road next to railroad. The area of interest lies immediately east of sand and gravel operation.

BEACON ROCK (W/4.8)   Approximately 800 ft.-high rock near MP 35 on SR-14, part of Beacon Rock State Park. Easy 0.9 mile trail just west of parking area leads to top. In winter, trailgate at MP 0.2 is locked.

BEACON ROCK POND (W/4.8)   1.5 acre pond in Beacon Rock State Park, about 1,500 ft. south of Beacon Rock. Shown as Ridell Lake on USGS Beacon Rock Quadrangle. Trail to pond starts across SR-14 from park headquarters.

BEACON ROCK SHORE (W/4.8)   Natural shoreline reached by Beacon Rock Moorage Road off SR-14 at MP 34.22; part of Beacon Rock State Park.

BEAR CREEK ROAD (W/7.0)   National Forest Road 6808 off Wind River Highway.

BELLE CENTER ROAD (W/2.4)   Turns north from SR-14 at MP 24.2.

BENSON PLATEAU (O/6.8)   4,000 ft. elev. plateau in Columbia Wilderness; reached by Pacific Crest Trail from Bridge of the Gods in Cascade Locks.

BERGE ROAD (W/7.6)   Paved road off SR-14 at MP 49.87, just west of Home Valley.

BIG HUCKLEBERRY MOUNTAIN (W/7.6)   4,200 ft. elev. point on Washington side of Gorge; reached by Pacific Crest Trail from Road 68 in Gifford Pinchot National Forest.

BIG LAVA BED (W/8.6)   Large (14,000 acres), gently sloping lava bed in Little White Salmon River drainage; bounded on east by South Prairie Road (Road 66); reached from MP 56.3 on SR-14.

BIGGS JUNCTION (O/19.0)   Small community at Exit 104 on I-84, at junction of I-84 with US-97.

BINGEN (W/11.5)   Community at MP 66.3 on SR-14, about 1.2 miles east of Hood River Bridge.

294

BINNS HILL ROAD (O/9.9)   Paved, then gravel, road climbing steeply west out of Oak Grove in Hood River Valley.

BLM PARCEL (O/17.6)   60 acres of public land managed by the Bureau of Land Management, lying along south side of Oregon Highway 206 about 0.8 mile east of Celilo Exit (#97) from I-84.

BOBS POINT (W/18.6)   Flat, rocky area along Columbia River; reached by gravel road (1.6 miles) from north end of US-97 bridge, north of Biggs Junction.

BONNEVILLE DAM (O & W/5.7)   Columbia River dam accessed by Exit 40 on I-84. The old navigation lock is about 0.5 mile from freeway, just south of the powerhouse.

BRADFORD ISLAND (O/5.8)   Island between Bonneville Dam and original powerhouse; specifically, shore southeast of parking area at Bonneville Dam Visitor Center.

BRIDAL VEIL (O/2.9)   Group of houses along Scenic Highway (MP 17); accessed by Exit 28 from I-84 (eastbound only).

BRIDAL VEIL CREEK (O/2.8)   Large stream west of Bridal Veil on Scenic Highway (MP 15.8).

BRIDAL VEIL FALLS STATE PARK (O/2.8)   MP 15.7 on Scenic Highway. Trails lead from parking area east to Bridal Veil Falls and north to bluff overlooking Columbia River.

BROUGHTON LUMBER MILL (W/10.3)   MP 61.85 on SR-14.

BROWER ROAD (O/2.7)   Left at MP 4.3 on Larch Mountain Highway; connects with Palmer Mill Road at Bridal Veil Creek.

BUCK CREEK ROAD (W/10.8)   Signed as Road B-1000; gravel continuation of paved Northwestern Lake Road, which turns off White Salmon River Road (SR-141) about 4.25 miles north of SR-14.

BUCK CREEK TRAILHEAD #1 (W/10.5)   Near MP 1.25 on Nestor Peak Road.

CAMPBELL CREEK (O/12.0)   At 1.1 miles on Rock Creek Road out of Mosier, take Proctor Road (left) down to crossing of Campbell Creek.

CAPE HORN (W/2.7)   700 ft. elev. point where SR-14 traverses a vertical cliff face (MP 25).

CAPE HORN ROAD (W/2.8)   Paved road south from SR-14 at MP 26.45; dead-ends at railroad tracks.

CARROLL ROAD (O/12.7)   Leads about 6 miles south from Mosier to junction with Vensel Road at 1,600 ft. elev.

CARSON DEPOT ROAD (W/7.2)   Unpaved road leading north from SR-14 at MP 47.91.

CASCADE LOCKS (O/6.3)   Community at south end of Bridge of the Gods, which spans the Columbia River; accessed by Exit 44 from I-84.

CATHERINE CREEK (W/12.7)   Small stream heading on Major Creek Plateau; crosses SR-14 at MP 72.2; crosses Old Highway 1.5 miles from west end; goes dry by mid to late summer.

CELILO (O/17.5)   Indian village just west of Celilo Exit (#97) on I-84.

CELILO PARK (O/17.5)   Beside Columbia River at Celilo Exit (#97) on I-84.

CHENOWETH ROAD (O/14.3)   Leave I-84 (eastbound) at Exit 82 (West The Dalles), go 0.6 mile west, past Wahtonka High School, to junction with Chenoweth Road (also "Chenowith" Road). Road climbs west to 2,100 ft. elev. to junction with Vensel Road; upper section is rather rough.

CHINIDERE MOUNTAIN (O/7.3)   4,673 ft. elev. peak near head of Eagle Creek; reached by Pacific Crest Trail from Wahtum Lake, or by gated road north of Wahtum Lake Forest Camp.

CLARK-SKAMANIA COUNTY LINE (W/2.0)   MP 21.7 on SR-14. Because of heavy traffic, best approached westbound where there is parking space for two or three cars. For safety, walk in ditch.

COLUMBIA HILLS (W/14.5-19.0)   2,200-3,200 ft. elev. ridge on north side of Columbia River opposite The Dalles; most easily accessed by The Dalles Mountain Road.

COLUMBIA WILDERNESS (O/5.4-8.5)   39,000 acre wilderness area in Mt. Hood National Forest; wilderness boundary lies about 1.5 miles south of Columbia River.

COOK (W/9.1)   Group of houses at MP 56.3 on SR-14.

COOK HILL (W/9.0)   3,000 ft. elev. point about 2 miles east of Dog Mountain. Dog Creek flows between Dog Mountain and Cook Hill.

CORBETT BOAT RAMP (O/1.6)   Access to Columbia River shore at Exit 22 from I-84.

CORBETT HILL ROAD (O/1.6)   Connects I-84 with Scenic Highway; accessed at Exit 22 from I-84.

CRATES POINT (O/14.4)   East end of Seven-mile Hill; take Rowena Exit (#76) and proceed east on US-30 to MP 12.9.

CRATES POINT DUNES (O/14.4)   Dunes between I-84 and railroad tracks. Accessed via I-84 (westbound) at MP 79.8; park on broad gravel shoulder beyond end of guardrail.

CRATES POINT WILDLIFE AREA (O/14.6)   Area of lakes between I-84 and Columbia River, just east of Crates Point; best approached by roads from the industrial zone of The Dalles.

CROWN POINT (O/2.1)   600 ft. vertical cliff south of I-84 at MP 24.7, directly across I-84 from Rooster Rock; accessed at top (Vista House) by Scenic Highway (MP 11.4).

DALLESPORT (W/14.8)   Small farming community accessed from US-197 just north of The Dalles Bridge, or from MP 82.25 on SR-14.

DALTON POINT (O/3.1)   Boat ramp allowing access to Columbia River shore; exit I-84 (westbound) at MP 29.

DEER MEADOW (W/4.3)   Large boggy meadow at elev. 3,100 ft.; head of east fork of Woodward Creek; accessed by rough Woodward Creek Road from MP 35 on SR-14.

DESCHUTES RIVER (O/18.1)   Accessed via Oregon Highway 206 from Celilo Exit (#97) from I-84.

DEVILS REST (O/3.5)   2,400 ft. elev. point 1.25 miles east of Angels Rest; reached by trail from Wahkeena Falls, or from Palmer Mill Road (Road 1520).

DNR FALLS (W/16.3)   Vernal waterfall on Washington Department of Natural Resources land extending up from SR-14 between MP 88.3 and MP 88.8. The eastern of the two falls is on state-owned land.

DNR FOREST (W/14.0)   560 acres of Washington Department of Natural Resources forestland and meadow at top of Fisher Hill, at junction of Fisher Hill Road and Timber Valley Road at 2,000 ft. elev.; includes all but southeast corner of Section 36; also reached via Lyle-Appleton Road and Timber Valley Road.

DODSON (O/4.6)   Group of houses along Scenic Highway (frontage road) near MP 24-25; accessed by Exit 35 from I-84.

DOG CREEK FALLS (W/9.0)   MP 56 on SR-14.

DOG MOUNTAIN (W/8.6)   3,000 ft. elev. point reached by Dog Mountain Trail from SR-14 at MP 53.8.

DRY CREEK ROAD (O/12.4)   Leads southeast from Mosier about 6 miles to junctions with State, Seven-mile, and Vensel (Osborn Cutoff) roads at about 1,400 ft. elev.

EAGLE CREEK (O/5.9)   Stream draining major part of Columbia Wilderness; flows into Columbia River near MP 41 on I-84.

EAGLE CREEK FOREST CAMP (O/5.9)   U.S. Forest Service campground near mouth of Eagle Creek; accessed by Exit 41 from I-84 eastbound.

EAGLE CREEK OVERLOOK PICNIC AREA (O/5.9)   Lies between I-84 and Columbia River; accessed by Exit 41 from I-84 eastbound.

EAGLE CREEK TRAIL (O/6.0)   13.3 mile trail along Eagle Creek, joining Pacific Crest Trail at Wahtum Lake; trailhead lies about 0.3 mile south of Eagle Creek Forest Camp.

EAST GORGE (O & W/10.7-19.0)   Columbia Gorge east of White Salmon River.

FIFTEEN MILE CREEK (O/15.5)   Stream flowing into Columbia River just below The Dalles Dam; mouth of creek lies about 500 ft. north of The Dalles Dam Visitor Center.

FISHER HILL GRADE (W/13.9)   Also called Fisher Hill Road; starts at bridge over Klickitat River at MP 1.7 on Klickitat River Road (SR-142).

FORT CASCADES HISTORIC SITE (W/5.5)   Turn south from SR-14 at MP 38.55; proceed 0.2 mile to parking area and information kiosk, from which trails traverse the historic site.

FRANZ LAKE (W/4.1)   Columbia River bottomland on Washington side about 3 miles west of Beacon Rock. Near MP 33.5 on SR-14, turn south on Skamania Landing Road, cross railroad tracks, take immediate right and proceed about 1 mile. As of this writing (1988), the area was owned by Trust for Public Land and was being proposed as a national wildlife refuge.

FULTON CANYON (O/18.4)   About 2 miles east of Deschutes River. Fulton Canyon is accessed by Oregon Highway 206 from either Celilo or Biggs Junction exits on I-84.

GAGING STATION (W/14.1)   Rocky area between SR-14 and Columbia River at MP 78.4.

GIBSON ROAD (W/1.5)   Turns northeast from SR-14 at MP 18.9.

GILLETTE LAKE (W/5.7)   4 acre lake along Pacific Crest Trail at about MP 3.0, measured from trailhead on SR-14 near Bridge of the Gods (MP 41.5).

GORGE TRAIL (O/0.4-10.1)   Low elevation trail parallel to Scenic Highway and I-84, extending from Sandy River at Lewis and Clark State Park to Hood River. As of 1988, several segments of the trail had not yet been built.

GOVERNMENT COVE (O/6.9)   Columbia River bay adjacent to I-84 near MP 47.

GOVERNMENT COVE QUARRY (O/6.9)   Partially quarried island about 2 miles east of Cascade Locks. From Cascade Locks, take frontage road to causeway leading to island.

GRANT LAKE (W/8.3)   11 acre lake on national forest land beside SR-14 at MP 53.1.

GRASSY KNOLL (W/7.9)   Large open area at 3,650 ft. elev. on ridge southeast of Big Huckleberry Mountain; reached by trail from Grassy Knoll Trailhead.

GRASSY KNOLL TRAILHEAD (W/8.1)   Large meadow area near junction of National Forest Roads 68 and 511, at 2,600 ft. elev.; trail to Grassy Knoll starts here.

HAINES ROAD (O/2.2)   Starts at MP 1.55 on Larch Mountain Highway; crosses Latourell Creek at about 0.7 mile and eventually joins Brower Road.

HAMILTON ISLAND (W/5.5)   Former island, now part of mainland downstream from Bonneville Dam and site for deposition of spoil from construction of second powerhouse. Western shores of the "island" are still in natural condition; accessed from MP 38.55 on SR-14.

HAMILTON MOUNTAIN (W/4.9)   2,445 ft. elev. point in Beacon Rock State Park; accessed by Hamilton Mountain Trail from park's picnic area, with a vertical gain of 2,000 ft. in about 3.5 miles. To reach picnic area, turn north from SR-14 just opposite Beacon Rock and drive 0.4 mile.

HARDY FALLS (W/4.9)   Waterfall at about MP 1.5 on Hamilton Mountain Trail.

HAYSTACK BUTTE (W/18.1)   2,969 ft. elev. point on Columbia Hills ridge opposite mouth of Deschutes River. Accessed by road from US-97 via Stringstreet and Dooley roads, but the last 2.5 miles are gated and must be traveled on foot.

HERMAN CREEK (O/6.9)   Stream draining part of Columbia Wilderness; enters Columbia River just east of Cascade Locks.

HERMAN CREEK ROAD (O/7.0-7.8)   South of I-84, extending about 3.3 miles from Exit 51 on I-84 to frontage road near Exit 47 (westbound).

HERMAN CREEK TRAIL (O/7.0)   11.2 mile trail from Columbia River to Pacific Crest Trail at 4,200 ft. elev. Trailhead at U.S. Forest Service Work Center on frontage road about 2 miles east of Cascade Locks.

298

HOME VALLEY   (W/7.7)   Small community at MP 50 on SR-14.

HOME VALLEY PARK   (W/7.7)   County riverside park at MP 50 on SR-14.

HOOD RIVER   (O/10.8)   City at MP 62-64 on I-84.

HOOD RIVER BRIDGE   (O & W/11.0)   Spans Columbia River from Exit 64 on I-84 to MP 65 on SR-14.

HOOD RIVER MOUNTAIN MEADOW   (O/11.4)   Large meadow on both sides of Hood River Mountain Road about 1 mile south of Old Dalles Road; elev. 1,800-2,000 ft.

HOOD RIVER MOUNTAIN ROAD   (O/11.4)   North-south road near ridgetop of Hood River Mountain at elev. 1,600-2,000 ft. From I-84 at Exit 64, take Oregon Highway 35 south about 0.5 mile, then left on Panorama Point Road about 2 miles, then left on Old Dalles Road about 2 miles to ridgetop, then right on Hood River Mountain Road.

HORSETAIL CREEK   (O/4.3)   Source of Horsetail Falls.

HORSETAIL FALLS   (O/4.3)   MP 22.2 on Scenic Highway, about 2.5 miles east of Multnomah Falls. Forest Service trail leads from base of falls to Upper Horsetail Falls and beyond.

HORSETHIEF BUTTE   (W/15.8)   Prominent rock formation in Horsethief Lake State Park. Loop trail leads to butte and around west side; trailhead at about MP 86.3 on SR-14.

HORSETHIEF LAKE STATE PARK   (W/15.7)   Includes Horsethief Butte, Horsethief Lake, and plateau area containing several vernal ponds. Turn off SR-14 at about MP 85.1.

I-84 (Interstate Highway 84)   Extends full length of Columbia Gorge near river level on the Oregon side.

ICE HOUSE LAKE   (W/6.2)   Small lake along SR-14 near Bridge of the Gods at MP 41.5. One of many pothole lakes formed as a result of ancient Table Mountain landslide.

INDIAN MOUNTAIN   (O/7.1)   4,880 ft. elev. peak near head of Eagle Creek; reached from Indian Springs by Pacific Crest Trail and Indian Mountain Trail.

INDIAN SPRINGS   (O/7.2)   Gently-sloping subalpine area (elev. 4,200 ft.) on north side of Indian Mountain; accessed by Indian Springs Road from Wahtum Lake Forest Camp, about 2.5 miles.

IVES ISLAND   (W/5.1)   About 2.5 miles below Bonneville Dam and a few hundred feet east of Pierce Island. In low water months (August and September) it can be reached on foot from west tip of Hamilton Island.

KLICKITAT FISHERMAN'S PARK   (W/14.2)   Public access site 4.85 miles up Klickitat River Road (SR-142).

KLICKITAT FISH LADDER   (W/13.8)   West shore of Klickitat River about 2.2 miles upstream from mouth; reached by unmarked dirt road at about MP 0.3 on gravel Fisher Hill Grade. The area is an Indian Fishing Site.

KLICKITAT RIVER   (W/13.5)   Enters Columbia River at MP 76 on SR-14. Paved Klickitat River Road (SR-142) follows east bank of river.

LARCH MOUNTAIN (O/3.9) Elev. 4,056 ft.; head of Multnomah and Oneonta creeks; accessed by Larch Mountain Highway (MP 14.5) and by a short trail to Sherrard Point viewpoint.

LARCH MOUNTAIN CORRIDOR (O/2.8-3.5) 4 mile protected forest corridor along Larch Mountain Highway from MP 4.7 to Mt. Hood National Forest boundary at MP 8.9.

LARCH MOUNTAIN HIGHWAY (O/2.0-4.0) Paved highway from Scenic Highway (MP 10.6) to top of Larch Mountain.

LATOURELL FALLS (O/2.3) MP 13.8 on Scenic Highway; loop trails lead to base of main falls and to base of upper falls.

LAWS CORNER (W/11.9) From White Salmon, take Snowden Road; near top of Major Creek Plateau, turn right on Bates Road and travel 1.8 miles to Laws Corner, 2,000 ft. elev. At this right-angle intersection of Bates Road and Bristol Road, the southwest corner is 480 acres of state-owned land open to foot travel by public.

LAWTON CREEK (W/1.8) MP 20.9 on SR-14.

LINDSEY CREEK (O/8.4) MP 53.7 on I-84.

LITTLE WHITE SALMON RIVER (W/9.3) MP 56.9 on SR-14; outlet of Drano Lake.

LITTLE WIND RIVER (W/7.5) Tributary of Wind River. A bridge spans Little Wind River at its mouth.

LOCKE LAKE (W/12.2) Lake(s) formed by railroad and highway causeways, which cut them off from Columbia River; MP 69.8 on SR-14.

LOWER TANNER CREEK ROAD (O/5.6) 0.3 mile dead-end road near the mouth of Tanner Creek. At Bonneville Exit (#40) from I-84, turn south on gravel road parallel to Tanner Creek. The road is gated, so the rest of this short road must be walked. From end of the road, a trail continues about a mile to Tanner Creek Falls.

LYLE (W/13.6) Community at mouth of Klickitat River; MP 76 on SR-14.

LYLE-APPLETON ROAD (W/13.5) Turns north from Old Highway (County Road 1230) about 1.25 miles out of Lyle.

LYLE CEMETERY (W/13.1) From SR-14 at MP 71, take Old Highway 3.1 miles, then left on Balch Road 0.3 mile.

LYLE TUNNEL (W/13.7) Highway tunnel on SR-14 just east of Lyle at MP 76.5.

McCLOSKEY CREEK ROAD (W/2.8) At MP 26.4 on SR-14, turn north on Salmon Falls Road, go about 0.8 mile to junction with Mabee Mines Road; after about 2 miles, turn right on rough McCloskey Creek Road.

McCORD CREEK FALLS (O/5.1) In Yeon State Park, MP 25.8 on Scenic Highway; from parking area, trails lead to base and top of falls.

MAJOR CREEK (W/12.8) Stream heading on Major Creek Plateau; crosses SR-14 at MP 72.6; crosses Old Highway about 2.2 miles from west end.

MAJOR CREEK PLATEAU (W/12.0-13.6) Large, 1,800-2,400 ft. elev. plateau at head of Major Creek, Catherine Creek, and Silvas Creek; take

Snowden Road out of White Salmon, Fisher Hill Road from MP 1.7 on Klickitat River Road, or Lyle-Appleton Road.

MAJOR CREEK ROAD   (W/12.8)   1.7 mile dead-end road which turns north from Old Highway 2.2 miles from west end. Best to park along Old Highway and walk the road, which has very little traffic.

MARYHILL MUSEUM   (W/18.7)   Art museum south of SR-14 near MP 99.

MAYER STATE PARK   (O/13.5)   US-30 east of Mosier at MP 6.8; also reached by Mayer Park Exit (#76) from I-84; referred to here is portion of Mayer State Park south of I-84, roughly 250 acres.

MEMALOOSE REST AREA   (O/12.8)   MP 73 on I-84 (eastbound).

MEMALOOSE STATE PARK   (O/12.8)   MP 73 on I-84 (westbound).

MEMALOOSE VIEWPOINT   (O/12.8)   State park land at MP 3.25 on US-30, east of Mosier.

MIDDLE GORGE   (O & W/7.5-13.5)   Loosely defined as region between Wind River and Klickitat River. Includes parts of west Gorge and east Gorge. A range given as "west and middle Gorge" includes west Gorge and eastward to Klickitat River. "East and middle Gorge" includes east Gorge and westward to Wind River.

MILLER ISLAND   (W/17.9-18.4)   Large (over 900 acres) island on Washington side of Columbia River channel, opposite mouth of Deschutes River.

MIRROR LAKE   (O/2.1)   Lake at base of Crown Point in Rooster Rock State Park, fed by Latourell Creek, drains to Columbia River. Accessed by dirt road off Exit 25 on I-84.

MITCHELL POINT   (O/9.6)   1,200 ft. elev. point along I-84 at MP 58.3; easily ascended from powerline corridor on south side. Access is by an exit from I-84 (eastbound); also trailhead for Wygant Trail.

MOFFETT CREEK   (O/5.2)   Flows into Columbia River at MP 39 on I-84.

MONTE CARLO   (W/10.2)   3,900 ft. elev. point on north boundary of Gorge, actually a 1 mile open ridge crest; head of Buck Creek; reached by Buck Creek Road (B-1000), then Road B-1800 to trailhead, then 2 mile trail to top.

MOSIER   (O/12.2)   Community at mouth of Mosier Creek. Accessed by Exit 69 on I-84.

MOSIER CREEK ROAD   (O/12.4)   Paved road just southeast of Mosier; runs along west side of Mosier Creek.

MOUNT DEFIANCE   (O/8.3)   Highest point in Gorge at 4,960 ft. elev.; reached by trail from Starvation Creek State Park, by trail from North Lake Trailhead, and from east by primitive jeep road, which is gated at Warren Lake Trailhead on Road 2821. Top of mountain is covered with electronic installations and has little botanical interest.

MT. HOOD NATIONAL FOREST BOUNDARY   (O/3.5)   MP 8.9 on Larch Mountain Highway; MP 19.0 on Scenic Highway. Mt. Hood National Forest lies east of these mileposts.

MT. ZION   (W/2.5)   1,466 ft. elev. point near Cape Horn. From SR-14 at MP 24.2, take Belle Center Road about 1.4 miles, then right on Strunk Road about 0.8 mile, again right on Mt. Zion Road about 0.6 mile to gate; walk to top.

MULTNOMAH FALLS   (O/3.6)   MP 30 on I-84; MP 19.7 on Scenic Highway.

MUNRA POINT   (O/5.5)   1,800 ft. elev. point south of I-84 at MP 39.5; top reached by primitive and somewhat dangerous trail up northwest ridge.

MURDOCK   (W/14.8)   Small community at MP 81.75 on SR-14.

NATIONAL FOREST ROAD 68   (W/6.6-8.1)   Provides access to Big Huckleberry Mountain Trail and Grassy Knoll Trail; reached by way of Wind River Highway (MP 47.44 on SR-14) and Panther Creek Road (Road 65).

NESTOR PEAK   (W/10.2)   3,088 ft. elev. point 6.5 miles north of Columbia River; reached by taking Buck Creek Road (B-1000) 0.4 mile, then left on Nestor Peak Road (N-1000) about 5 miles, then a trail on right leads 1 mile to top of Nestor Peak.

NORTH FORK WASHOUGAL RIVER ROAD   (W/2.5)   Turns north from Washougal River Road near MP 15.

NORTH LAKE   (O/7.9)   Small lake at 4,000 ft. elev. in Columbia Wilderness.

NORTH LAKE TRAILHEAD   (O/8.2)   About 4,000 ft. elev. on Road 2820 out of Dee in Hood River Valley; serves as trailhead for North Lake, Bear Lake, and Mt. Defiance; spring and marshy area are nearby.

NORTHWESTERN LAKE   (W/10.6)   Reservoir behind Condit Dam on White Salmon River; turn left from White Salmon River Road about 4.2 miles from SR-14.

OLD HIGHWAY   (W/12.4-13.6)   Also called County Road 1230; section of old Evergreen Highway, bypassed some years ago by present river-level SR-14; extends from Rowland Lake to Klickitat River. Turn off SR-14 at MP 71, or just west of Klickitat River near MP 76.

OLD US-30   (O/11.0-12.1)   Old Columbia River Highway between Hood River and Mosier. Two sections, both dead-end at this time: west section starts from Highway 35 at east city limits of Hood River, proceeding east about 4 miles to a dead end; east section starts from about MP 0.7 on Rock Creek Road out of Mosier and proceeds west about 0.6 mile to a dead end. Current (1988) Highway Department plans call for linking these two segments of old Columbia River Highway to create a bicycle and pedestrian pathway between Hood River and Mosier.

ONEONTA BRIDGE   (O/4.1)   Old Columbia River Highway bridge over Oneonta Creek at MP 22 on Scenic Highway.

ONEONTA GORGE   (O/4.1)   Narrow gorge of Oneonta Creek at MP 22 on Scenic Highway.

OREGON HIGHWAY 206   (O/17.5-19.0)   Runs parallel to I-84 between Celilo Exit (#97) and Biggs Junction Exit (#104).

PACIFIC CREST TRAIL (PCT)   North-south trail mostly on national forest lands and generally following crest of Cascade Mountains. Washington and Oregon trailheads are respectively at north and south ends of Bridge of the Gods at Cascade Locks.

PALMER MILL ROAD (O/2.9)   Steep road parallel to Bridal Veil Creek, climbing 2,500 ft. from Bridal Veil on Scenic Highway (MP 16.3) to junction at MP 9.9 on Larch Mountain Highway.

PIERCE ISLAND (W/4.9)   Nature Conservancy property close to Beacon Rock.

PIERCE NATIONAL WILDLIFE REFUGE (W/5.0)   Approximately 300 acre refuge between Beacon Rock and the town of North Bonneville; primarily a refuge for geese; not open to general public visitation at this time (1988).

PRINDLE MOUNTAIN (W/3.2)   Elev. 1,800 ft.; take Salmon Falls Road from MP 26.4 on SR-14; at top of hill, take Ryan-Tavelli Road. Several old logging roads traverse much of the wooded top of Prindle Mountain.

RAINY LAKE (O/7.8)   Elev. 4,100 ft.; lies close to Rainy Lake Road (Road 2820), which starts at Dee in Hood River Valley. After crossing bridge at Dee, turn right.

RIDGE TRAIL (O/2.3)   Approximately 3 mile loop trail which starts at entrance to Rooster Rock State Park. Trail runs east along wooded ridge parallel to I-84 and returns west to parking area near park office.

ROCK CREEK LAKE (W/6.3)   West of Stevenson; originally part of Columbia River at mouth of Rock Creek; cut off from river by causeway for railroad and SR-14.

ROCK CREEK ROAD (O/12.1)   Southwest of Mosier; starts at Mosier Exit (#69) from I-84; first paved, then gravel, finally primitive dirt road intersecting Hood River Mountain Road at 1,600 ft. elev.

ROOSTER ROCK (O/2.0)   Pointed rock formation on north side of I-84 at MP 24.3; part of Rooster Rock State Park; accessed from west end of parking lot.

ROOSTER ROCK STATE PARK (O/1.8-2.8)   On both sides of I-84 between MP 24 and MP 28; includes Sand Island.

ROWENA (O/13.8)   Small community on US-30; accessed from I-84 at Exit 76.

ROWENA CREST (O/13.4)   Top of Rowena Plateau (elev. 800 ft.) in Mayer State Park (MP 7 on US-30).

ROWLAND LAKE (W/12.4)   East of Bingen at MP 71 on SR-14; originally part of Columbia River, but cut off by railroad and highway causeways; public access on north side of lake.

RUCKEL CREEK (O/6.0)   Small stream 0.5 mile east of Eagle Creek; best accessed by Gorge Trail from Eagle Creek Forest Camp.

SAND ISLAND (O/2.6)   Part of Rooster Rock State Park; accessible on foot at low water, generally August and September.

SANDY RIVER (O/0.4)   MP 18 on I-84; west boundary of Columbia Gorge.

SCENIC HIGHWAY (O/0.7-5.0)   26 mile segment of old (1915) Columbia River Highway, from Sandy River to McCord Creek; accessed at Corbett (#22), Bridal Veil (#28), and Dodson (#35).

SEVEN-MILE HILL   (O/13.3-14.3)   2,000 ft. elev. east-west ridge west of The Dalles.

SEVEN-MILE HILL ROAD   (O/14.2)   Parallel to Seven-mile Hill, but 1.5-2 miles to the south; turns right from Chenoweth Road about 0.5 mile west of The Dalles; reaches an elevation of 1,800 ft., where it connects with Dry Creek Road and other roads.

SHELLROCK MOUNTAIN   (O/8.2)   2,100 ft. elev. point along I-84 at MP 52, roughly opposite Wind Mountain on Washington side.

SHEPPERD'S DELL   (O/2.6)   State park at MP 15 on Scenic Highway, west of Bridal Veil; short trail to waterfall on Young Creek.

SHERRARD POINT   (O/4.0)   Viewpoint on Larch Mountain, elev. 4,056 ft.; reached by quarter-mile trail from parking lot.

SIGNAL ROCK   (W/7.0)   Viewpoint about 185 ft. above Columbia River east of Stevenson. At MP 46.7 on SR-14, take easternmost of 2 dirt roads.

SILVAS CREEK   (W/13.8)   Small stream heading on Major Creek Plateau; flows into Klickitat River about 1.5 miles above its mouth.

SILVER STAR MOUNTAIN   (W/2.2)   4,400 ft. elev. peak on northwest boundary of Gorge; reached from Washougal on the south or from East Fork Lewis River Road on the north, but last few miles to top are primitive roads which are closed to vehicles and must be walked.

SKYLINE ROAD   (O/14.5)   Scenic road starting on Mt. Hood Street in The Dalles, climbing to 2,000 ft. elev. in about 7 miles, and remaining at 2,000 ft. elev. for several miles on Dutch Flat.

SMITH-CRIPE ROAD   (W/3.6)   Turns north from SR-14 near MP 29.5.

SNOWDEN ROAD   (W/11.3)   Runs north and east out of White Salmon, climbing to Major Creek Plateau at 2,000 ft. elev.; paved most of way to Snowden.

SOUTH PRAIRIE ROAD   (W/9.4)   National Forest Road 66; lies along east margin of Big Lava Bed; reached by taking road from Cook (MP 56.3 on SR-14) to Willard; Road 66 starts about 1 mile north of mill at Willard.

SPEARFISH LAKE   (W/15.4)   Approximately 20 acre lake in 50 acre Spearfish Lake Park, administered by U.S. Army Corps of Engineers; reached by taking road east from US-197, 0.5 mile north of The Dalles Bridge; road dead-ends at Spearfish Lake parking area in about 1 mile.

SPRING CAMP ROAD   (O/3.9)   0.3 mile primitive road off Larch Mountain Highway at MP 11.6; best to walk it; Larch Mountain Trail 441 crosses road at its upper end.

SPRING CREEK FISH HATCHERY   (W/10.3)   Lies between railroad and Columbia River at MP 61.5 on SR-14.

SR-14 (State Route 14)   Principal highway on Washington side of Columbia River.

STANLEY ROCK   (O/11.3)   Part of Koberg Beach State Park, about 1.5 miles east of Hood River at MP 66 on I-84 (westbound).

STARVATION CREEK STATE PARK (O/8.8)   MP 55 on I-84 (eastbound); short trail to Starvation Creek Falls; trailhead for Mt. Defiance and Warren Creek trails.

STEVENS POND (O/12.8)   Vernal pond. From Mosier, take State Road 2 miles; turn right on Morgensen Road, park; pond lies 200 ft. to west.

STEVENSON (W/6.4)   Town along SR-14 at MP 44-45; county seat of Skamania County.

TABLE MOUNTAIN (W/5.2)   3,400 ft. elev. peak north of Bonneville Dam; accessed by Pacific Crest Trail and connecting trails.

TANNER CREEK TRAIL (O/5.7)   Trail 431 in Columbia Wilderness; accessed by 5 mile powerline road starting at Bonneville Exit (#40) from I-84. Not to be confused with Tanner Creek Falls Trail from end of Lower Tanner Creek Road, dead-ending at Tanner Creek Falls.

THE DALLES AIRPORT (W/15.0)   An area of roughly 1,200 acres on north side of Columbia River; accessed from US-197 just north of The Dalles Bridge, or from SR-14 near Murdock via Dallesport Road.

THE DALLES BRIDGE PONDS (W/15.3)   Several ponds and swales lying west of US-197 about 100 yards from north end of The Dalles Bridge. A portion is U.S.-owned (Corps of Engineers); the rest is part of The Dalles Airport.

THE DALLES BRIDGE ROAD (W/15.3)   US-197 between SR-14 and The Dalles Bridge; accessed by Exit 87 from I-84.

THE DALLES DAM VISITOR CENTER (O/15.5)   Take Exit 87 from I-84; turn left on US-197; at frontage road, turn right and drive east about 0.9 mile to parking area.

THE DALLES MOUNTAIN ROAD (W/15.4)   Gravel road turning north from SR-14 at MP 84.5; crosses Columbia Hills ridge crest at 2,200 ft. elev., 8 miles from SR-14.

THE DALLES RIVERSIDE PARK (O/15.2)   Columbia River shoreline accessed via Exit 85 from I-84.

THREE CORNER ROCK (W/4.4)   3,500 ft. elev. point about 8 miles north of Beacon Rock; accessed by Washougal River Road and Stebbins Creek Trail.

THREE CREEK CAMP (O/7.3)   Trail campsite 11 miles up Eagle Creek Trail from Eagle Creek Trailhead, or 2.5 miles down from Wahtum Lake Forest Camp.

TOMLIKE MOUNTAIN (O/7.4)   4,500 ft. elev. peak between east and west forks of Herman Creek.

TOM McCALL NATURE PRESERVE (O/13.4)   231 acre Nature Conservancy area along US-30 out of Mosier at about MP 7; adjacent to Mayer State Park.

TUNNELS 1, 2, 3, 4, & 5 (W/9.5-10.0)   Series of tunnels on SR-14 east of Dog Mountain, between MP 58.2 and 60.2.

US-30 (O/12.2-14.5)   Old Columbia River Highway between Mosier and The Dalles.

VENSEL ROAD (O/12.8)   Upper elevation road southeast of Mosier; connects with Dry Creek, Carroll, and Chenoweth roads.

VIENTO ROCKS (O/9.1)   Prominent rock formations along I-84 at Viento State Park.

VIENTO STATE PARK (O/9.0)   Exit 56 from I-84.

WAHKEENA CREEK (O/3.5)   From Wahkeena Falls, creek flows into Wahkeena Pond and out again and is joined by Multnomah Creek (outlet of Benson Lake). Wahkeena/Multnomah Creek flows under I-84 and into Columbia River near MP 30.5 on I-84.

WAHKEENA FALLS (O/3.5)   MP 19.2 on Scenic Highway, about 0.5 mile west of Multnomah Falls.

WAHTUM LAKE (O/7.5)   Head of East Fork of Eagle Creek; elev. 3,800 ft.; accessed by 0.25 mile trail from Wahtum Lake Forest Camp, which is reached by a paved road out of Dee in Hood River Valley.

WARREN LAKE TRAILHEAD (O/8.7)   At 3,800 ft. elev., northeast of Mt. Defiance; trailhead is on National Forest Road 2821, a branch of Road 2820, out of Dee in Hood River Valley.

WASCO BUTTE (O/13.0)   Elev. 2,346 ft. Best accessed out of Mosier by Carroll and Vensel roads. Spur road leads to top, where a lookout and several electronic facilities are located. A small pond lies just below top on north side.

WASHOUGAL (W/0.8)   Small city at about MP 16 on SR-14.

WASHOUGAL RIVER ROAD (W/0.8)   Paved road (SR-140) starting in Washougal and following river upstream, eventually changing to good gravel road (W-2000).

WEST FORK WASHOUGAL RIVER BRIDGE (W/2.1)   At MP 6.7 on Washougal River Road (SR-140), turn left on Bear Prairie Road 3.3 miles, then right on Skamania Mines Road (W-1240) about 1.5 miles to West Fork Washougal River.

WEST GORGE (O & W/0.4-10.7)   Columbia Gorge west of White Salmon River.

WHITE SALMON (W/11.2)   Town along SR-141 about 1 mile from Bingen.

WHITE SALMON RIVER (W/10.7)   Large stream entering Columbia River at MP 63.6 on SR-14, about 1.5 miles west of Hood River Bridge.

WHITE SALMON RIVER ROAD (W/10.7)   Paved road on east side of White Salmon River from MP 63.6 on SR-14. At about 1.75 miles, it joins SR-141 and continues north.

WILLARD ROAD (W/9.1)   Take Cook-Underwood Road from MP 56.3 on SR-14 to road fork at around 5 miles; left fork leads north to Willard.

WIND MOUNTAIN (W/7.9)   1,900 ft. elev. peak at MP 51-52 on SR-14.

WIND RIVER (W/7.5)   Major stream which enters Columbia River at about MP 49.5 on SR-14.

WIND RIVER HIGHWAY (W/7.2)   Turns north from SR-14 at MP 47.44;

306

becomes National Forest Road 30 at Gifford Pinchot National Forest boundary.

WIND RIVER SHORE   (W/7.5)   At Home Valley, take Berge Road, then Indian Cabin Road to bridge over Little Wind River. Follow Wind River shore upstream. A fee may be charged for use of this private land.

WISHRAM   (W/17.4)   Community at MP 92.7 on SR-14.

WISHRAM HISTORICAL MARKER   (W/17.6)   MP 93.6 on SR-14; commemorates Celilo Falls.

WYETH TRAIL   (O/7.8)   National Forest Trail 411; 6.2 mile trail which climbs to 4,000 ft. elev. at Road 2820; trailhead is at south end of Wyeth Forest Camp; accessed from I-84 at Exit 51.

WYGANT TRAIL   (O/9.5)   Oregon State Park trail originating at exit at MP 58.3 from I-84 (eastbound); trail runs west to Perham Creek and ascends 2,214 ft. elev. Wygant Point.

YEON STATE PARK   (O/5.0)   Includes McCord Creek Falls; parking area at MP 25.8 on Scenic Highway (Dodson frontage road).

YOUNG CREEK   (O/2.3-2.7)   Crosses Scenic Highway at Shepperd's Dell (MP 15); flows eastward in lowlands of Rooster Rock State Park, joining Latourell Creek just east of Mirror Lake.

BIBLIOGRAPHY

Abrams, L. *An Illustrated Flora of the Pacific States, Washington, Oregon, and California.* 4 vols. Palo Alto: Stanford University Press, 1940-1960.

Clark, L. G. *Wild Flowers of British Columbia.* Sidney, B.C., Canada: Gray's Publishing Limited, 1973.

Detling, L. E. "Peculiarities of the Columbia River Gorge Flora." *Madroño 14* (1958): 160-172.

Franklin, J. F., and C. T. Dyrness. *Natural Vegetation of Oregon and Washington.* Portland: USDA Forest Service General Technical Report PNW-8, 1973.

Haskin, L. L. *Wild Flowers of the Pacific Coast.* Portland: Binfords & Mort, 1959.

Hitchcock, C. L., and A. Cronquist. *Flora of the Pacific Northwest.* Seattle: University of Washington Press, 1974.

Jolley, R., and L. Kemp. *Survey of Wildflowers and Flowering Shrubs of the Columbia Gorge.* Portland: Native Plant Society of Oregon, 1984.

Meinke, R. J. *Threatened and Endangered Vascular Plants of Oregon: An Illustrated Guide.* Portland: U.S. Fish and Wildlife Service, 1982.

Oregon Natural Heritage Data Base. *Rare, Threatened and Endangered Plants and Animals of Oregon.* Portland: The Nature Conservancy, 1987.

Peck, M. E. *A Manual of the Higher Plants of Oregon.* Portland: Binfords & Mort, 1941.

Rickett, H. W. *The Northwestern States.* Vol. 5 of *Wild Flowers of the United States.* New York: McGraw-Hill, 1971.

Siddall, J. L., K. L. Chambers, and D. H. Wagner. *Rare, Threatened and Endangered Vascular Plants in Oregon.* Salem, Ore.: Oregon State Land Board, 1979.

Washington Natural Heritage Program. *An Illustrated Guide to the Endangered, Threatened and Sensitive Vascular Plants of Washington.* Olympia, Wash.: Washington Natural Heritage Program, 1981.

Washington Natural Heritage Program. *Endangered, Threatened & Sensitive Vascular Plants of Washington.* Olympia, Wash.: Washington State Department of Natural Resources, 1987.

WILDFLOWER INDEX

THE NUMBERS listed in this index are picture numbers and refer the reader to the numbered entries in the Guide section.

312

314

315

318

FIELD NOTES

COLOPHON

TWO TYPEFACES are used in *Wildflowers of the Columbia Gorge*. Syntax, a sans serif typeface noted for its readability, is used for display type and technical text. Trump Mediaeval, a serif typeface designed by Georg Trump, is the text face. *Wildflowers of the Columbia Gorge* is printed on 70 lb. Sterling dull.

Production of this volume was accomplished through the skill and cooperation of the following:

TYPESETTING:	Irish Setter
DISPLAY ORNAMENTS:	Mackenzie-Harris Corp.
COLOR SEPARATIONS:	Portland Prep Center, Inc.
PRINTING:	Publishers Press
EDITING:	Krisell Steingraber
DESIGN:	George Resch